ENERGY SCIENCE, ENGINEERING AND TECHNOLOGY

BIOFUEL USE IN THE U.S.

IMPACT AND CHALLENGES

Energy Science, Engineering and Technology

Additional books in this series can be found on Nova's website under the Series tab.

Additional E-books in this series can be found on Nova's website under the E-books tab.

ENERGY SCIENCE, ENGINEERING AND TECHNOLOGY

BIOFUEL USE IN THE U.S.

IMPACT AND CHALLENGES

Stefania Alonso
and
Maria Regina Ortega
EDITORS

Nova Science Publishers, Inc.
New York

For permission to use material from this book please contact us:
Telephone 631-231-7269; Fax 631-231-8175
Web Site: http://www.novapublishers.com

NOTICE TO THE READER

Additional color graphics may be available in the e-book version of this book.

Library of Congress Cataloging-in-Publication Data

Biofuel use in the U.S. : impact and challenges / [edited by] Stefania Alonso and Maria Regina Ortega.
p. cm.
Includes index.
ISBN 978-1-62100-441-7 (hardcover)
1. Biomass energy--United States. 2. Biomass energy industries--United States. 3. Energy policy--United States. I. Alonso, Stefania. II. Ortega, Maria Regina.
HD9502.5.B543U6125 2011
333.95'390973--dc23

2011034017

Published by Nova Science Publishers, Inc. † New York

Contents

Preface

Achieving greater energy security by reducing dependence on foreign petroleum is a goal of U.S. energy policy. The Energy Independence and Security Act of 2007 (EISA) calls for a Renewable Fuel Standard (RFS-2), which mandates that the United States increase the volume of biofuel that is blended into transportation fuel from 9 billion gallons in 2008 to 36 billion gallons by 2022. Long-term technological advances are needed to meet this mandate. This book examines how meeting the RFS-2 would affect various key components of the U.S. economy. If biofuel production advances with cost-reducing technology and petroleum prices continue to rise as projected, the RFS-2 could provide economy-wide benefits. However, the actual level of benefits to the U.S. economy depends importantly on future oil prices and whether tax credits are retained in 2022.

Chapter 1- Diversifying the Nation's energy supply is one of the primary means for providing long-term energy security. A diverse energy portfolio can also have far-reaching economic impacts by reducing dependence on foreign oil. The Energy Independence and Security Act of 2007 (EISA) mandates a Renewable Fuel Standard (RFS-2) under which the United States will annually produce 36 billion gallons of biofuel, primarily ethanol, by 2022. Transitioning away from nonrenewable fossil fuels (such as petroleum oil) without placing additional burden on the U.S. economy is a long-term challenge. Although experts and policymakers generally agree on the importance of energy security, how best to achieve this goal and at what cost is subject to debate.

Chapter 2- U.S. transportation relies largely on oil for fuel. Biofuels can be an alternative to oil and are produced from renewable sources, like corn. In 2005, Congress created the Renewable Fuel Standard (RFS), which requires transportation fuel to contain 36 billion gallons of biofuels by 2022. The most common U.S. biofuel is ethanol, typically produced from corn in the Midwest, transported by rail, and blended with gasoline as E10 (10 percent ethanol). Use of intermediate blends, such as E15 (15 percent ethanol), would increase the amount of ethanol used in transportation fuel to meet the RFS. The Environmental Protection Agency (EPA) recently allowed E15 for use with certain automobiles.

Chapter 3- The market for biomass-based diesel (BBD) fuel, most notably biodiesel, has expanded rapidly since 2004, largely driven by federal policies, especially tax credits and a mandate for their use under the federal Renewable Fuel Standard (RFS). Most expect that the majority of the BBD fuel quota in the RFS will be met using biodiesel produced from soybean oil. Biodiesel from other feedstocks, and other biomass-based substitutes (e.g., synthetic diesel from cellulosic feedstocks or algae) could play a larger role in the future, although currently these other alternatives are prohibitively expensive to produce in sufficient quantities.

Chapter 4- In the search for renewable fuel alternatives, biofuels have gained strong political momentum. In the last decade, extensive mandates, policies, and subsidies have been adopted to foster the development of a biofuels industry in the United States. The Biofuels Initiative in the Mississippi Delta resulted in a 47-percent decrease in cotton acreage with a concurrent 288-percent increase in corn acreage in 2007. Because corn uses 80 percent more water for irrigation than cotton, and more nitrogen fertilizer is recommended for corn cultivation than for cotton, this widespread shift in crop type has implications for water quantity and water quality in the Delta. Increased water use for corn is accelerating water-level declines in the Mississippi River Valley alluvial aquifer at a time when conservation is being encouraged because of concerns about sustainability of the groundwater resource. Results from a mathematical model calibrated to existing conditions in the Delta indicate that increased fertilizer application on corn also likely will increase the extent of nitrate-nitrogen movement into the alluvial aquifer. Preliminary estimates based on surface-water modeling results indicate that higher application rates of nitrogen increase the nitrogen exported from the Yazoo River Basin to the Mississippi River by about 7 percent. Thus, the shift from cotton to corn may further contribute to hypoxic (low dissolved oxygen) conditions in the Gulf of Mexico.

In: Biofuel Use in the U.S.
Editors: S. Alonso and M. R. Ortega
ISBN: 978-1-62100-441-7

Chapter 1

EFFECTS OF INCREASED BIOFUELS ON THE U.S. ECONOMY IN 2022

Mark Gehlhar, Ashley Winston, and Agapi Somwaru

ABSTRACT

Achieving greater energy security by reducing dependence on foreign petroleum is a goal of U.S. energy policy. The Energy Independence and Security Act of 2007 (EISA) calls for a Renewable Fuel Standard (RFS-2), which mandates that the United States increase the volume of biofuel that is blended into transportation fuel from 9 billion gallons in 2008 to 36 billion gallons by 2022. Long-term technological advances are needed to meet this mandate. This report examines how meeting the RFS-2 would affect various key components of the U.S. economy. If biofuel production advances with cost-reducing technology and petroleum prices continue to rise as projected, the RFS-2 could provide economywide benefits. However, the actual level of benefits (or costs) to the U.S. economy depends importantly on future oil prices and whether tax credits are retained in 2022. If oil prices stabilize or decline from current levels and tax credits are retained, then benefits to the economy would diminish.

Keywords: Bioenergy, economywide, ethanol, petroleum, trade, macro-economic factors, RFS-2

ACKNOWLEDGMENTS

The authors thank the following individuals for their valuable insights and recommendations: Peter Dixon, Maureen Rimmer, Hosein Shapouri, and Wally Tyner.

SUMMARY

Diversifying the Nation's energy supply is one of the primary means for providing long-term energy security. A diverse energy portfolio can also have far-reaching economic impacts by reducing dependence on foreign oil. The Energy Independence and Security Act of 2007 (EISA) mandates a Renewable Fuel Standard (RFS-2) under which the United States will annually produce 36 billion gallons of biofuel, primarily ethanol, by 2022. Transitioning away from nonrenewable fossil fuels (such as petroleum oil) without placing additional burden on the U.S. economy is a long-term challenge. Although experts and policymakers generally agree on the importance of energy security, how best to achieve this goal and at what cost is subject to debate.

What Is the Issue?

Reducing dependence on foreign energy by expanding domestic renewable fuels can have impacts for the overall U.S. economy because of energy's importance in consumption, production, and trade. In the past, increasing energy independence would generally be expected to place a greater burden on the U.S. economy because of the higher domestic costs of producing alternative energy to replace relatively inexpensive foreign petroleum. However, according to the U.S. Department of Energy, petroleum prices are more likely to continue rising in the long term relative to the cost of producing domestic biofuels. Although the exact timing is uncertain, cost-reducing technology in biofuel production is expected to be a key factor in expanding production and making biofuels competitive with petroleum. However, without policies that provide incentives to deploy renewable energy technology, biofuel producers likely will shy away from investing in new technology because of market uncertainty. The RFS-2 mandate is accompanied by incentives in the form of tax credits to ethanol blenders. Tax credits, however, could add to taxpayers' costs and place greater burden on the economy. This study examines the potential effects of the RFS-2 on the U.S. economy as measured by gross domestic product (GDP), household income and consumption, price and quantity of energy fuels, and agricultural production and trade. We compare the U.S. economy in 2022 with and without the RFS-2.

What Did the Study Find?

If biofuel production technology advances and petroleum prices continue to rise as projected, the RFS-2 could benefit the U.S. economy. U.S. household consumption would rise because of higher real wages, increased household income, and lower import prices. By substituting domestic biofuels for imported petroleum, the United States would pay less for imports overall and receive higher prices for exports, providing a gain for the economy from favorable terms of trade. Improved technology and increased investment would enhance the ability of the U.S. economy to expand.

Gross Domestic Product

Changes in GDP and the magnitude of benefits (or costs) are highly dependent upon assumptions about alternative biofuel support policies and the future price of oil. The greater the value of displaced petroleum for each dollar of biofuel produced and the lower the tax credits, the greater the benefit to the U.S. economy. Cost-reducing technology would not only reduce the costs of producing biofuels but also contribute to national GDP because production would rise as efficiency improves.

Household Welfare

Household consumption would increase regardless of whether or not tax credits were retained, with the gains primarily due to increased real income, favorable terms of trade with relatively lower import prices, and hence, greater purchasing power to the household. Consumption would increase by about $13-$28 billion, depending largely on oil prices. The RFS-2 would raise real wages and household disposable income as returns to labor and capital increase. Replacing imported oil with domestic biofuels would lower the cost of motor fuels. Thus, households would spend less on, but consume more, motor fuels. In addition, lower prices for imports and fuel would encourage greater consumption of other goods and services.

Energy and Trade

Expansion of domestic biofuel production would reduce petroleum demand, thereby reducing the quantity of imported crude petroleum. Crude oil, which is a major input for gasoline, would be displaced by ethanol. U.S. imports of crude oil would fall by 16-17 percent in 2022. The United States is the largest importer of crude oil, with imports accounting for about two-thirds of total U.S. supply. Reduced U.S. demand for petroleum would lower the price of crude oil. As a result of lower demand and a decline in the import price, the U.S. import bill for crude oil would decline by $61-$68 billion. With a smaller import bill, the U.S. dollar would appreciate. A stronger dollar would reduce the cost of importing other goods, including agricultural commodities, and reduce export volume because of increased prices in foreign markets for U.S. products. In addition, with greater demand for land to use for both energy crop production and all other agricultural activities, meeting the RFS-2 would reduce U.S. agricultural commodity exports and increase the demand for agricultural imports as crops must compete for limited land.

Caveats

This study does not predict the future but addresses the question of what would be the likely impacts on the U.S. economy should the RFS-2 mandate be met under different price/policy scenarios. The study acknowledges the uncertainty in meeting the mandate in 2022. The exact timing of the commercialization of new technologies to produce biofuels cannot be determined because of a myriad of uncertain factors. Future developments will also depend on new investments in infrastructure needed to support a transportation and distribution network for biofuels. Determining when such developments would take place is beyond the scope of this study. Long-term impacts on the U.S. economy from meeting the RFS-2 will depend partly on future petroleum prices. This study adopts a price projection from the U.S. Department of Energy that assumes that satisfying the growing world demand for petroleum will require accessing higher cost supplies of oil. Under these conditions, petroleum prices are likely to be higher in 2022 than current prices. Unlike previous decades, petroleum prices are likely to rise relative to the cost of producing biofuels.

Note: In real prices.

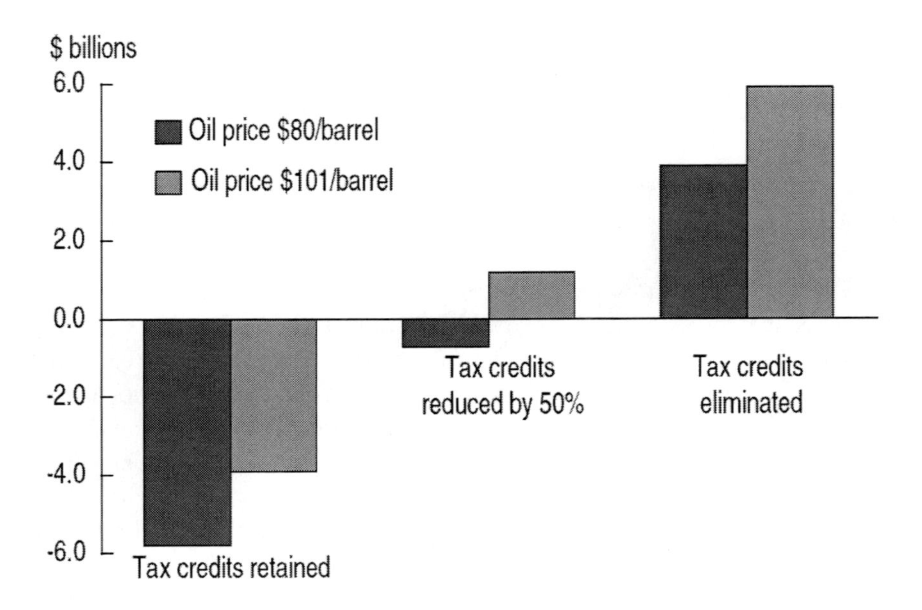

Impact of RFS-2 on U.S. gross domestic product.

How Was the Study Conducted?

The study used a detailed computable general equilibrium model for the United States—the U.S. Applied General Equilibrium (USAGE) model—comprising 534 industries. The model is a multipurpose framework for addressing a broad set of questions, including domestic and trade policy as well as macroeconomic links to trade. The model was modified to include additional sectors and industries involved in biofuel production, including conventional ethanol (corn-starch) produced from dry-milling and second-generation ethanol made from crop residues, dedicated energy crops, and other advanced biofuels. Other modifications include explicit treatment of U.S. agricultural land and regional land allocation for the production of biomass (organic material) and all other agricultural activities. A base, or reference, scenario without the RFS-2 was conducted for the year 2022 using the U.S. Department of Energy's projections. The effects of RFS-2 were determined as alternative scenarios using scenario analysis. Volumes of all types of ethanol were based on those established by EISA.

INTRODUCTION

Diversifying the Nation's energy supply is the primary means for reducing long-term energy dependence on foreign sources. Although experts and policymakers generally agree on the importance of reducing dependence on imported petroleum, how best to achieve this goal is debatable. The Energy Independence and Security Act of 2007 (EISA) calls for a Renewable Fuel Standard (RFS-2), which mandates that the United States increase the volume of biofuel that is blended into transportation fuels from nearly 9 billion gallons in 2008 to 36 billion gallons by 2022. The RFS-2 mandate is accompanied by incentives in the form of tax credits, which are subject to change over time, to ethanol blenders. This study assesses the long-term effects of the RFS-2 by projecting into 2022 its impact on U.S. gross domestic product (GDP), trade, and household welfare.

Success in meeting the mandate depends on overcoming various challenges. One challenge is to diversify the sources of ethanol feedstocks beyond conventional corn, which involves greater reliance on cost-reducing technology, particularly if ethanol is to compete with petroleum in the future. A second challenge is to ensure sufficient demand for biofuels. Demand for transportation fuels in the United States is expected to continue to decrease as fuel efficiency standards for vehicles are raised. A third challenge is to find a solution to the "blend wall" constraint, in which increased ethanol production meets the 10-percent ethanol-to-gasoline blending limit. Regardless of the cost competitiveness of ethanol, the blend constraint could technically halt ethanol's growth unless domestic use is increased by altering automotive engine requirements to accommodate higher ethanol blends.

The effect on the U.S. economy in meeting the RFS-2 could depend on the incentives used to encourage investment. Without policies that provide incentives, underinvestment to advance renewable energy technology is likely due to risk aversion from market uncertainty (Rajagopal et al., 2009). Some argue that tax credits could be justified on the basis of achieving national energy security (Tyner, 2007). Others argue in favor of providing temporary protection to an infant industry, where learning by doing and adapting new technologies have future payoffs (Sheldon and Roberts, 2008). Although policies may be needed to correct market failures, sustaining incentives indefinitely may not be beneficial as industries mature. Economic incentives in the form of tax credits could add to taxpayers' costs and place greater burden on the economy. Some argue that industries with a mature technology, such as corn ethanol production, may no longer need tax credit incentives.[1] This report considers implications of tax credits on the U.S. economy in 2022.

Another factor that can affect the impact of the RFS-2 on the U.S. economy is the competitiveness of biofuels. In the past, biofuels have not been competitive with petroleum oil. However, with cost-reducing technology, the prospects improve for biofuels to become competitive. In this case, biofuels might compete with imported petroleum oil in the future. Based on the U.S. Department of Energy's (DOE) (2009) long-term projections, ethanol prices in real terms are expected to fall, whereas petroleum-based gasoline prices would continually rise as crude oil prices rise with growing world demand. If DOE's projections hold, they may present an opportunity for the United States to reduce its reliance on tax credits as ethanol becomes more competitive with petroleum. This study determines the impacts of the RFS-2 based on the DOE's price forecast.

The long-term gains to the economy from using cost-reducing technology in the production of ethanol, especially advanced biofuels, could offset costs associated with tax credits. The RFS-2 requires that ethanol be produced from a broad variety of organic material (biomass), with production of cellulosic ethanol (made from wood, grasses, algae, or nonedible parts of plants) increased from that of the previous mandate. Second-generation biofuels, such as cellulosic ethanol, have yet to exploit commercially existing technology advances. The farm and energy sectors are connected in many ways that can have important implications for meeting the RFS-2. The type of biomass used to satisfy the mandate can affect regional land allocation, farm production, and commodity trade.

In this study, we examine the interactions among energy and other sectors of the economy and estimate the long-term effects of the RFS-2 on the U.S. economy—that is, GDP, trade, and household welfare (income and consumption). Although there is much momentum for producing cellulosic ethanol, this study neither predicts when the technology for producing cellulosic ethanol will be adopted for commercial use nor determines the feasibility of meeting the timetable set by the 2007 EISA.

However, the study does present scenarios that take into account projected energy prices based on DOE long-term forecasts and tax credits for different types of ethanol. The study demonstrates that the RFS-2 mandate, when met under different conditions, involves tradeoffs for the U.S. economy.

SCENARIO ALTERNATIVES

To analyze the effects of the RFS-2, we compare the U.S. economy with and without the mandate in the year 2022. This comparison allows us to ascertain the impact of only the RFS-2, holding other factors constant. The projection of the U.S. economy in 2022 without RFS-2 is called the *reference* scenario (see appendix 1). Alternative scenarios of the U.S. economy with the RFS-2 assume full implementation of the mandate for the year 2022. However, it is unknown whether the RFS-2 will be met by retaining current tax credits for the duration of the mandate. Because the impact of the RFS-2 could depend on tax credits and petroleum oil prices, we present alternative scenarios with different tax credits and oil prices (see box, "Scenarios and Energy Price Projections"). We adopt the projected prices determined by DOE's long-term forecast.

The combination of different oil price assumptions, a low price (LP) of $80 dollars per barrel, a high price (HP) of $101 per barrel, and three tax credit assumptions provide six scenarios (S) to draw upon for the analysis:

(1) S1-LP: tax credits retained and low oil price.
(2) S2-LP: 50-percent reduction in tax credits and low oil price.
(3) S3-LP: no tax credits and low oil price.
(4) S1-HP: tax credits retained and high oil price.
(5) S2-HP: 50-percent reduction in tax credits and high oil price.
(6) S3-HP: no tax credits and high oil price.

SCENARIOS AND ENERGY PRICE PROJECTIONS

Long-term crude oil price projections depend on changing but highly unpredictable global petroleum supply and demand conditions. A number of factors may influence future prices of petroleum oil. For example, decisions made by the Organization of the Petroleum Exporting Countries (OPEC) can significantly affect global crude oil supply. DOE provides long-term forecasts of crude petroleum prices, supply, and use. We adopt DOE's forecast because projecting or forecasting petroleum prices is beyond of the scope of this report. According to DOE's reference forecast that assumes normal world growth, the price of crude oil is expected to gradually rise to $111 per barrel by the year 2030. The demand for petroleum oil would increase primarily from steady growth in developing countries (see appendix 2). Under normal growth conditions, the price of crude oil is projected to reach about $101 per barrel in 2022.

DOE projects the wholesale price of ethanol to remain near $2 per gallon through 2022. According to DOE, production costs are expected to fall in the long term, lowering the wholesale price of ethanol to below $2 per gallon through 2030. Improved efficiency in producing advanced ethanol, which would make ethanol more competitive with petroleum, is a key factor for the widening price gap between ethanol and gasoline over the next two decades. For this study, we adopt DOE's price for 2022 of $2.12 per gallon for ethanol in all scenarios, including scenarios that reduce or eliminate tax credits.

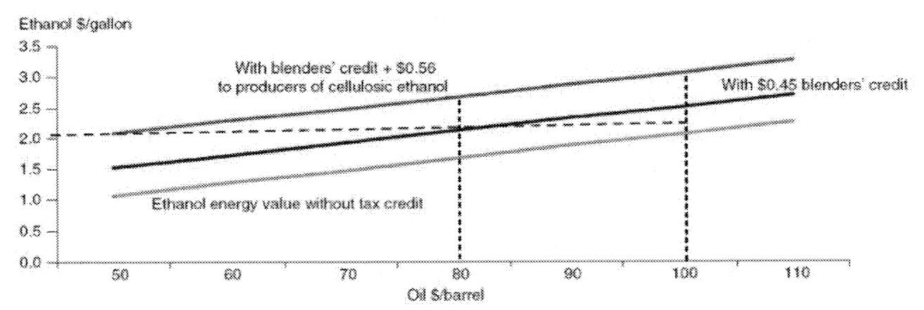

Source: Authors' calculations based on U.S. Department of Energy price projections. Energy price equivalent values.

In our scenarios, we use DOE's projected oil price for 2022 of $101 as a high price and a corresponding energy-equivalent oil price using DOE's projected ethanol price of $2.12 per gallon. The figure below illustrates the correspondence between petroleum oil and ethanol prices on an energy-equivalent basis.[1] An oil price of $80 per barrel would correspond to DOE's projected ethanol price of $2.12 per gallon in 2022 on an energy-equivalent basis. For example, blenders could be willing to pay up to $2.12 per gallon for ethanol on an energy-equivalent basis when the price of crude oil is $80 per barrel and when the blender receives the 45-cent tax credit currently paid for corn ethanol. With oil prices above $80 per barrel, blenders could afford to pay more than $2.12 per gallon for ethanol. With the oil price of $101 per barrel and the tax credit retained at 45-cents per gallon, blenders could afford to pay $2.52 per gallon for ethanol. If ethanol suppliers were to receive $1.01 per gallon, blenders could offer up to $3.08 per gallon for cellulosic ethanol. Depending on future costs and technology advancement, this price could profitably compensate cellulosic producers.[2]

[1] This assumes ethanol contains two-thirds the energy of gasoline and a constant price relationship between petroleum oil and gasoline we adopt from Tyner and Taheripour (2007).

[2] The cost of producing cellulosic ethanol will depend on the size of the biorefinery, which could range from 20 to 100 million gallons per year. The production cost for producing cellulosic ethanol ranges from $1.60 to $2.00 per gallon, with yields ranging from 76 to 88 gallons per dry ton, depending on feedstock material. For further details, see National Academies Press (2009).

QUANTIFYING THE EFFECTS OF THE RFS-2 ON THE U.S. ECONOMY

This section reports the results from the series of scenarios for the RFS-2. Results are simulated from an economywide modeling framework (see box, "The USAGE Model"). In doing so, the analysis identifies key links and interactions between energy, trade, and the farm sector from an economy-wide perspective.

Macroeconomic Impacts

Macroeconomic factors, such as GDP, household income and consumption, and international trade, are the broadest impact measures on the national economy. Meeting the RFS-2 results in multiple long-term effects on the U.S. economy. One effect is from the substitution of ethanol for gasoline in motor fuels. Replacing petroleum gasoline with less expensive ethanol would reduce domestic spending on motor fuels. This reduction would occur as long as ethanol is competitive with imported petroleum oil. The cost savings from reduced motor fuel expenditure are passed on to households. A second effect is from the reallocation of factors of production (production inputs). Industries with relatively high rates of technological progress generate increased returns to factors employed, thereby raising income to households. The overall economy benefits from resource allocation as returns to labor and capital increase. A third effect is from changes in trade and international prices for exports and imports. Household welfare increases when export prices rise relative to import prices (known as a "terms of trade effect").

This study measures gains or losses from changes in GDP and household welfare (in real terms).[2] The larger the value of displaced petroleum for each dollar of biomass produced, the greater the benefit to the U.S. economy. The higher the projected oil price, the greater the increase in GDP and household welfare. With oil at $80 per barrel and tax credits fully retained (S 1-LP), GDP would decrease by $5.8 billion, partly because of the forgone revenue from tax credits. However, with oil at $101 per barrel and tax credits fully retained (S1-HP), GDP would fall a little less, by $3.9 billion (figure 1).

Cost-reducing technology plays an important role in meeting the RFS-2 not only by reducing ethanol's costs but also by contributing to national GDP. Assuming that reductions in tax credits are coupled with greater technological advances, lower tax credits would provide additional benefits to the economy. Reducing tax credits by 50 percent would increase GDP by $1.2 billion under the high oil price scenario. However, eliminating tax credits could raise U.S. output by nearly $6 billion under the high oil price scenario and by almost $4 billion under the low oil price scenario. Technological improvements in biofuel production could offset the negative effects associated with tax credits by improving efficiency and raising output in meeting the mandate. The contribution from technology would vary, depending on the price for oil (table 1). Under the assumption that technological progress permits ethanol of all types to become competitive with petroleum, the U.S. economy would benefit in meeting the RFS-2 as imported crude oil is reduced. The higher the U.S. import bill is for petroleum, the greater the potential for GDP to increase in meeting the RFS-2.

The USAGE Model

This study uses a model known as the United States Applied General Equilibrium model (USAGE). The model is an economywide framework of the U.S. economy and is designed for projections and policy analysis. USAGE is a multipurpose model developed by the Centre of Policy Studies at Monash University in collaboration with the U.S. International Trade Commission (USITC). The original theoretical basis for the model is based on Dixon and Rimmer (2002). The agricultural cost component is constructed from the Agricultural Resource Management Survey (ARMS) developed by USDA's Economic Research Service.

Several modifications were made for this application by adding additional industries, which provide explicit treatment for distinct types of biomass material for ethanol production and use (Winston, 2009). Modifications were necessary to represent implementation of the RFS-2, which specifies that ethanol be produced domestically from different sources, and to link agricultural production to regional land categories. One of the significant modifications to the USAGE model is the individual treatment of different types of ethanol using different feedstock materials.[1] This modification was necessary in order to implement the specified volumetric requirements of the RFS-2, which calls for different amounts of ethanol from three sources: corn ethanol, cellulosic ethanol, and other advanced biofuels.

Decisions for changes in land allocation are determined by producers' profit- maximizing behavior. The model includes biomass material—including cornstarch, cellulosic (crop residues and perennial energy crops), and other advanced (primarily from forestry residue)—for producing different types of ethanol. Differences in tax credits are also taken into account for individual types of ethanol. The model allows for flexibility in the substitution of multiple inputs for production of different types of ethanol (see appendix 3).

[1] Previous versions of the USAGE model did not distinguish different types of ethanol and biomass.

Household consumption would increase in all scenarios, and the consumption gains would be higher than GDP gains (figure 2). Increases in consumption would depend on crude oil prices, tax credit reductions, and technology advances for meeting DOE's price projections. With oil at $80 per barrel and tax credits fully retained (S 1-LP), consumption would increase by about $13 billion. With oil at $101 per barrel (S3-HP), consumption would increase by $28 billion from meeting the RFS-2. Consumption gains are primarily the result of increased real income, lower import prices, and higher export prices, all of which lead to greater purchasing power to the household.

Returns to factors of production in the form of wages and returns to capital would be enhanced by replacing more expensive imported petroleum (table 1). Real wages and real household disposable income would increase from meeting the RFS-2. With oil at $80 per barrel and tax credits fully retained (S 1-LP), household disposable income would increase by 0.10 percent, but if tax credits were eliminated (S3-LP), income would increase by 0.15 percent (table 1). A higher oil price ($101 per barrel) with tax credits retained (S 1-HP) would raise household disposable income by 0.14 percent. Similarly, real wages for consumers would increase with reduced tax credits and a higher oil price (S3-HP), while the dollar would appreciate in real terms.

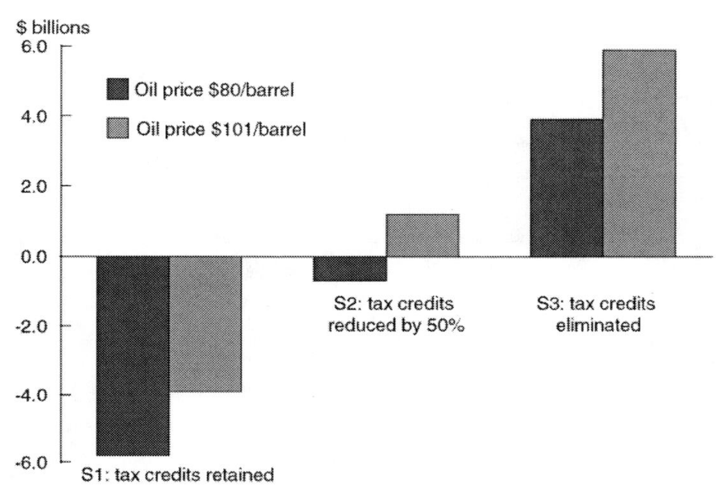

RFS-2 = Renewable Fuel Standard.
Source: USAGE model simulation.

Figure 1. Impact of RFS-2 on U.S. gross domestic product.

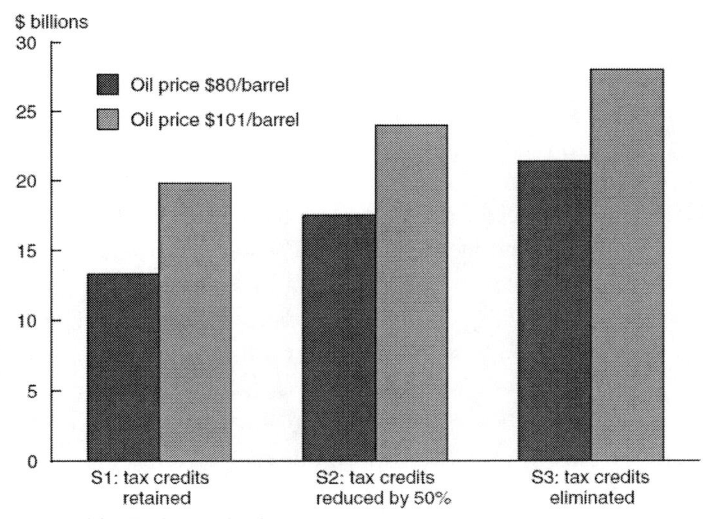

RFS-2 = Renewable Fuel Standard.
Source: USAGE model simulation.

Figure 2. Impact of RFS-2 on U.S. household consumption.

Table 1. Macroeconomic impact of RFS-2 in 2022[1]

Macroeconomic factors	S1-LP	S2-LP	S3-LP	S1-HP	S2-HP	S3-HP
	Percent change					
Household disposable income	0.10	0.12	0.15	0.14	0.17	0.20
Real wages for consumers	0.40	0.40	0.40	0.49	0.49	0.49
Real appreciation	0.91	0.94	0.96	1.16	1.19	1.21
Terms of trade	0.80	0.81	0.81	1 .00	1 .01	1 .02
Export price	0.09	0.08	0.07	0.09	0.08	0.07
Import price	-0.70	-0.72	-0.73	-0.89	-0.91	-0.93
Import volume	-0.02	-0.04	-0.06	0.08	0.05	0.04
Export volume	-0.93	-0.96	-0.98	-1.05	-1.08	-1.10
Returns to factors of production	0.307	0.310	0.313	0.378	0.381	0.384
Technology contribution (to GDP growth)	0.244	0.261	0.274	0.240	0.262	0.271

[1] 2022 reference base year.

RFS-2 = Renewable Fuel Standard.

S1: tax credits retained; LP: Low oil price, HP: High oil price. S2: tax credits reduced by 50 percent; LP: Low oil price; HP: High oil price. S3: tax credits eliminated; LP: Low oil price; HP: High oil price.

Source: USAGE model simulation.

Replacing higher priced imported oil with domestic biofuels yields greater benefits as consumer spending on motor fuels is reduced. Among all consumer expenditures, those for motor fuels would fall the most, which in turn would encourage greater spending on other goods and services. The impact on expenditures for motor fuels would depend on the oil price. Expenditures for motor fuels would fall by about $9 billion with oil at $80 per barrel and by about $11 billion with oil at $101 per barrel. In addition, under the high oil price scenario, returns to factors of production would increase. Because household income is derived from wages and returns on capital, disposable income would also be greater. Higher incomes would also increase consumption of other goods and services.

Energy Fuel Impacts

The impacts on the U.S. economy in meeting the RFS-2 would come partly from lower priced imported petroleum oil and reductions in the price of domestic fuel. Table 2 presents the effects of the RFS-2 on price and quantity of energy fuels. The United States, as the largest single-country importer of crude oil, could potentially lower the world price of crude oil by reducing its imports. Reduced U.S. demand for petroleum would lower both the price and quantity Joyce Newton

238 Asharoken Avenue

Northport, NY 11768

States as import prices would fall relative to export prices. The price of oil would fall about 4 percent. The reduction in the price of oil would also depend on future supply and demand conditions in the rest of the world.

Expansion of domestic biofuel production would reduce petroleum demand but increase the quantity demanded for motor fuels as the price of motor fuels falls (see box, "The RFS-2 and Implications for the Price and Quantity of Motor Fuels").[4] The price of motor fuels is indirectly affected by the price of crude petroleum. Crude oil is a major input for petroleum-based gasoline, which affects the cost of gasoline. The price of motor fuels (a blended price) would fall as the price of petroleum-based gasoline decreases.

The RFS-2 and Implications for the Price and Quantity of Motor Fuels

The effect of the RFS-2 on motor fuels plays an important role in determining economywide impacts. The impacts of the RFS-2 depend partly on the price and quantity changes for blended motor fuels and petroleum-based gasoline. Meeting the RFS-2 could result in either an increase or a decrease in the price of motor fuels, depending on supply assumptions for gasoline and ethanol. Our price projections are based on the DOE's long-term price forecast for petroleum oil and ethanol. According to this projection, increased ethanol production over the long term could take place without the price of ethanol rising in real terms. To impose the DOE's price projection in our model, the ethanol industry's technological change is treated as endogenous. In other words, the model solves for technological change needed to satisfy the given quantity change according to the RFS-2 while imposing the longrun price projection. In this setting, meeting the RFS-2 is not binding as the supply increases to meet the DOE price forecast. In meeting the RFS-2 in 2022, the prices of gasoline and motor fuels are endogenous. The supply of gasoline is determined by the supply of crude petroleum oil, which depends on assumed supply and demand conditions in the rest of the world. Technological change is constant for gasoline and motor fuels production as prices for these fuels change.

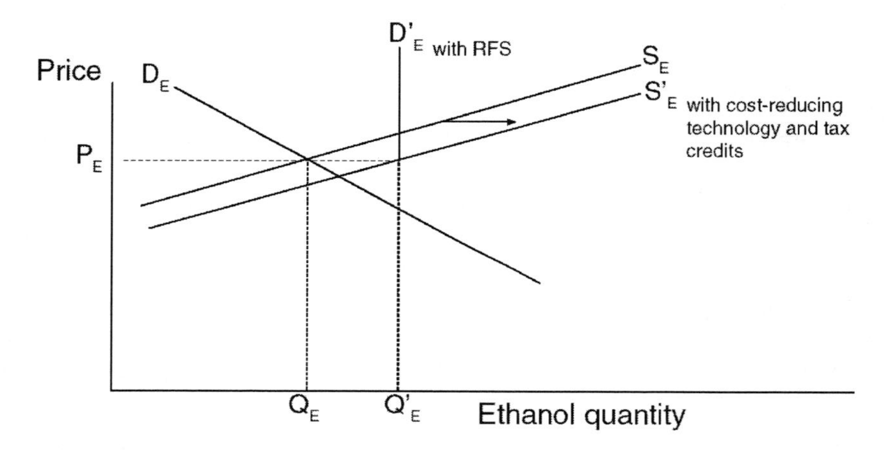

RFS achieved with cost-reducing technology in ethanol production.

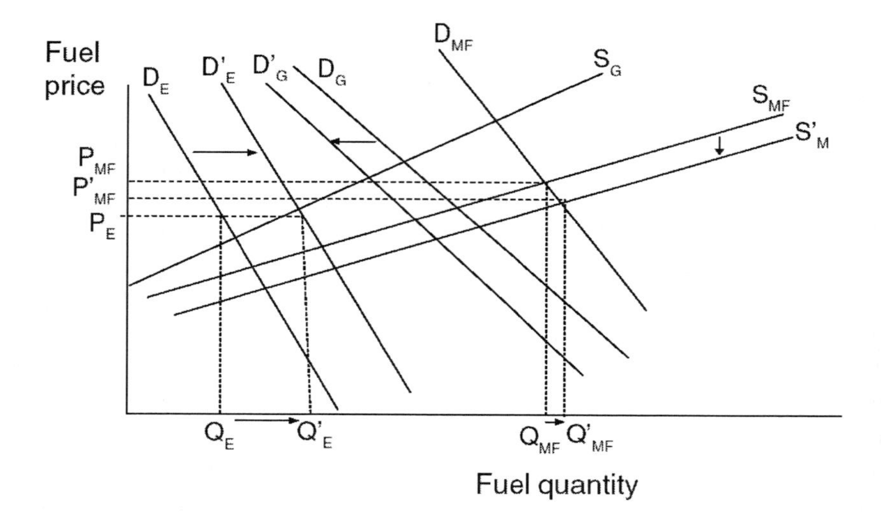

Longrun impact on price of motor fuels with reduced demand for gasoline.

The change in the equilibrium prices and quantities are depicted graphically in the accompanying diagrams. Meeting the requirement of the RFS-2 shifts the demand curve for ethanol from D_E to D'_E (top diagram). Cost- reducing technical change is denoted as a rightward shift in the supply of ethanol from S_E to S'_E . The quantity of ethanol increases from Q_E to Q'_E, while the price of ethanol P_E remains constant.

In this case the mandate is not binding and the market is able to satisfy the quantity demanded by RFS-2 without increasing the market price for ethanol. The price is determined from long-term reductions in the cost of producing ethanol, particularly for advanced ethanol.

Motor fuels, a mixture of gasoline and ethanol, get an increasing proportion of ethanol in meeting the RFS-2. The mixture of gasoline and ethanol depends on how much motor fuel quantity changes, which depends on the price of gasoline and ethanol. Ethanol displaces gasoline, thereby reducing gasoline demand, which is denoted as a shift from D_G to D'_G (bottom diagram). Given that the supply curve for gasoline (S_G) is upward sloping, the price of gasoline falls. The price of motor fuels falls as a result of substituting ethanol for gasoline and from using lower priced gasoline. The reduction in the price of gasoline allows the motor fuel industry to sell a greater quantity, shown as a rightward shift in the supply of motor fuels S_{MF}. Demand for motor fuels is depicted as a downward sloping curve D_{MF}. Consumers of motor fuels demand a higher quantity of motor fuels. The equilibrium quantity supplied and demanded for motor fuels increases from Q_{MF} to $Q'_{MF,}$ as the price of motor fuels fall from P_{MF} to $P'_{MF.}$ Total expenditures on motor fuels fall as a result of the drop in the price of motor fuels. The quantity demanded will depend on how responsive consumers are to a given change in the price of motor fuels.

[1] Previous versions of the USAGE model did not distinguish different types of ethanol and biomass.

The price of gasoline would fall by about 9 percent under the low oil price scenario and by about 8 percent under the high oil price scenario (table 2). The 12-percent drop in the price of motor fuels is largely the result of substituting lower cost ethanol for gasoline. The ethanol price would remain at $2.12 per gallon according to the projection we adopt from DOE for the year 2022. As consumption of motor fuels (blended) increases, gasoline use increases. Thus, in meeting the RFS-2, U.S. gasoline consumption would not be reduced by the same amount of gasoline replaced by ethanol.[5]

The value of imported crude oil would fall by $61 billion under the low oil price scenarios and by $68 billion under the high oil price scenarios (table 3). The total value of U.S. imports of all goods would fall under each scenario but would fall more with reduced tax credits and under the high price scenario for crude oil because the cost of importing goods would fall with an appreciating dollar. Non-oil imports would increase but would increase less with reductions in tax credits.

Table 2. Impact of RFS-2 on Domestic Fuel and Imported Crude Oil in 2022[1]

Item	Unit	S1-LP	S2-LP	S3-LP	S1-HP	S2-HP	S3-HP
	Percent change						
Crude oil imports	Price	-4.11	-4.13	-4.15	-3.77	-3.79	-3.82
	Quantity	-17.46	-17.51	-17.55	-15.96	-16.00	-16.05
Gasoline	Price	-8.71	-8.72	-8.78	-7.82	-7.83	-7.83
	Quantity	-20.31	-20.31	-20.31	-18.13	-18.13	-18.13
Motor fuels (blended gasoline)	Price	-11.73	-11.74	-11.74	-12.02	-12.02	-12.02
	Quantity	2.47	2.47	2.47	2.53	2.53	2.54

[1] 2022 reference base year.

RFS-2 = Renewable Fuel Standard.

S1: tax credits retained; LP: Low oil price, HP: High oil price. S2: tax credits reduced by 50 percent; LP: Low oil price; HP: High oil price. S3: tax credits eliminated; LP: Low oil price; HP: High oil price.

Source: USAGE model simulation.

Table 3. Impact of RFS-2 on U.S. Oil and Non-Oil Imports in 2022[1]

Oil/non-oil imports	S1-LP	S2-LP	S3-LP	S1-HP	S2-HP	S3-HP
	Change in $ billions					
Crude oil imports	-61.23	-61.42	-61.59	-67.52	-67.75	-67.95
Non-oil imports	20.48	18.22	16.29	21.05	18.79	16.88
Total imports	-40.75	-43.20	-45.30	-46.47	-48.95	-51.07

[1] 2022 reference base year.

RFS-2 = Renewable Fuel Standard.

S1: tax credits retained; LP: Low oil price, HP: High oil price. S2: tax credits reduced by 50 percent; LP: Low oil price; HP: High oil price. S3: tax credits eliminated; LP: Low oil price; HP: High oil price.

Source: USAGE model simulation.

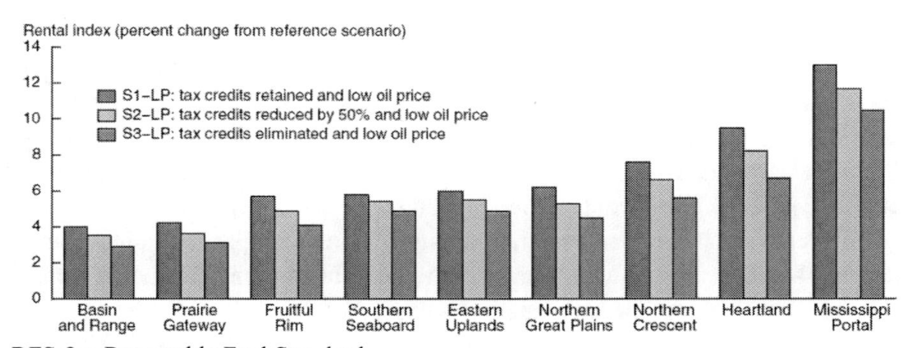

RFS-2 = Renewable Fuel Standard.
Source: USAGE model simulation.

Figure 3. Impact of RFS-2 on regional land returns, 2022.

Farm Production and Trade Impacts

As previously discussed, meeting the RFS-2 mandate would reduce fuel prices and contribute to the dollar's appreciation. Both of these factors have the effect of lowering costs for the farm and food sector. However, competition for limited land could place upward pressure on land rents. This analysis considers the returns to land by location and production allocation to crop and livestock activities. Producers tend to allocate land to production activities that are expected to yield the largest benefit. The expected returns to land depend on the price of outputs and inputs, producers' preferences, and land quality (Lubowski et al., 2006). If meeting the mandate under the RFS-2 could be accompanied by cost-reducing technology for biomass and ethanol production, the land requirement to produce energy crops would be lower, which in turn, would mean less competition for land reallocation.

The analysis assumes that producers would not make alterations to enhance land capability, such as capital investment in irrigation, land reclamation, or conversions that might enable crops to grow on acreage previously classified as not suitable for growing crops. With fixed acreage, land rents would be a reflection solely of reallocating land to activities with higher returns. Demands on land differ because of the regional differences in acreage availability for growing energy crops and production characteristics of other crop and livestock activities. Producers alter acreage toward more profitable activities.

Increases in dedicated energy crop production would increase competition for all land types and consequently would raise rental rates. The effect of RFS-

2 on land rents would depend on the proportion that energy crops would contribute to total cellulosic requirements. The magnitude of the increase in rental rates by region would capture differences in regional competition for land. Land types capable of supporting energy crops, including the Mississippi Portal, Heartland, and Eastern Uplands, are more likely to see higher rents (figure 3). These areas would experience higher increases in land rents because of the land suitability for accommodating different crops, including energy crops. The land rents would also rise in the Fruitful Rim and Basin and Range, but the increase would be smaller.

The abundance of farmland is not as important as its suitability for crop growing. For example, the Basin and Range region, with abundant and low-cost farmland, would be mostly unsuitable for growing energy crops due to inadequate rainfall and poor soil conditions. On the other hand, the Fruitful Rim region, with its suitable climate and soil conditions, would provide an ideal location for energy crops. However, energy crops would face greater competition with other traditional crops in this region. Land rents for rice would increase more than rents for other crops because of the indirect competition between rice and energy crop production (figure 4). Because live-stock and horticultural crops are produced in many places throughout the country, production of those commodities may face less intense competition from energy crops.

The RFS-2 would impact food prices considerably less than it would farm commodity prices in the long term (2022). For example, corn and rice prices would increase more than prices for downstream products, such as prepared feeds, for which prices increase by less than 0.5 percent (table 4). Farm commodity prices would rise partly because of increased competition for land, thereby raising production costs. Prices would increase less under alternative scenarios where tax credits are reduced.

The effects of the RFS-2 on crop production would vary, depending on the particular demand for each crop (table 5). Production of corn, used partly as an energy crop, would increase by 7-11 percent, whereas production of other crops not directly used for energy would likely fall. Improvements in cost-reducing technology, coupled with reduced tax credits, would reduce the effect of the RFS-2 on production under each scenario (S 1-LP through S3-HP). Technological gains made for cellulosic energy crops could reduce the effects on production for all crop and livestock sectors.

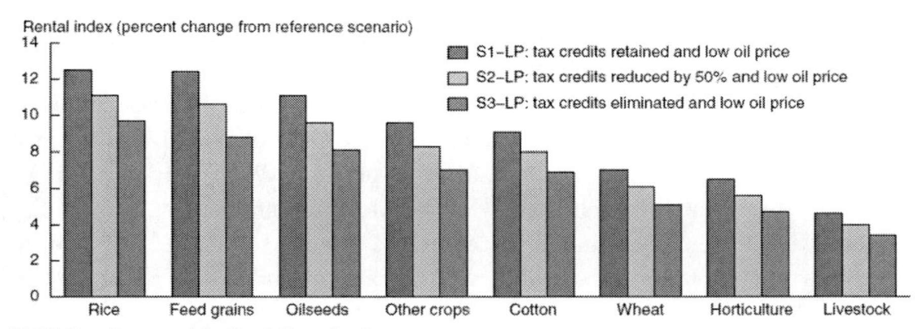

RFS-2 = Renewable Fuel Standard.
Source: USAGE model simulation.

Figure 4. Impact of RFS-2 on agricultural sector land rents, 2022.

Table 4. Impact of RFS-2 on Selected U.S. Farm and Food Prices in 2022[1]

Commodity	S1-LP	S2-LP	S3-LP	S1-HP	S2-HP	S3-HP
	Percent change					
Corn	4.81	4.06	3.23	4.68	3.94	3.12
Rice	2.03	1 .81	1 .59	1 .71	1 .50	1 .30
Fruits	0.71	0.62	0.52	0.63	0.54	0.45
Tree nuts	0.26	0.23	0.21	0.23	0.20	0.18
Meat	0.48	0.42	0.37	0.47	0.42	0.37
Fluid milk	0.52	0.46	0.40	0.53	0.48	0.42
Cheese	0.43	0.38	0.33	0.43	0.38	0.33
Flour	0.51	0.46	0.40	0.48	0.43	0.37
Prepared feeds	0.48	0.41	0.35	0.47	0.41	0.34
Bread	0.15	0.14	0.14	0.17	0.17	0.16

[1] 2022 reference base year.
RFS-2 = Renewable Fuel Standard.
S1: tax credits retained; LP: Low oil price, HP: High oil price. S2: tax credits reduced by 50 percent; LP: Low oil price; HP: High oil price. S3: tax credits eliminated; LP: Low oil price; HP: High oil price.
Source: USAGE model simulation.

Total land availability is constrained in this analysis. Thus, meeting the RFS-2 would reduce U.S. agricultural commodity exports and increase imports (tables 6 and 7) because domestic crops must compete with limited land. However, with gains from technological advances, the impacts on trade for all commodities would be reduced. Corn export and import flows would be affected more than those of other commodities by the RFS-2, with exports decreasing by 3-4 percent and imports rising 9-13 percent. For other commodities, reduced production would limit exports and increase imports. For example, with lower rice production, rice imports would rise by 3-4 percent, whereas exports would fall 1-2 percent. With imports further weakened by improved technology for cellulosic ethanol, most commodity imports would rise by less than 2 percent (table 7).

Increased biomass production would raise demand for farm inputs purchased from both domestic and foreign sources. This analysis assumes no capacity constraints in the United States or the rest of the world for farm input supplies. The responsiveness of farm input supply would moderate the impact of the RFS-2 on farm production and trade. Increased imports of farm inputs and lower energy prices would play an important role in determining domestic input supply response. Lower energy prices would reduce costs for domestic manufacturing of farm inputs. Both domestic fertilizer production and imported fertilizer would increase to meet increased demand (table 8).

Table 5. Impact of RFS-2 on U.S. Farm Output in 2022[1]

Commodity	S1-LP	S2-LP	S3-LP	S1-HP	S2-HP	S3-HP
	Percent change					
Corn	11.64	9.56	7.25	11.62	9.54	7.24
Rice	-1.43	-1.34	-1.25	-1.47	-1 .37	-1.28
Cotton	-1.18	-1.17	-1.17	-1.15	-1.14	-1.12
Wheat	-0.65	-0.64	-0.64	-0.63	-0.63	-0.62
Other feed grains	-0.06	-0.10	-0.14	-0.05	-0.09	-0.14
Soybeans	-0.46	-0.55	-0.66	-0.47	-0.56	-0.66
Other oilseeds	-0.74	-0.77	-0.82	-0.72	-0.75	-0.79
Dairy	-0.39	-0.33	-0.27	-0.43	-0.37	-0.30
Beef	-0.53	-0.49	-0.45	-0.54	-0.50	-0.46
Other livestock	-1 .01	-0.93	-0.84	-1 .00	-0.92	-0.83
Tobacco	-1.25	-1.19	-1.13	-1.22	-1.16	-1.10
Fruits	-1.38	-1.25	-1.11	-1.45	-1.31	-1.17
Tree nuts	-1 .71	-1 .63	-1 .55	-1 .65	-1 .56	-1 .47
Vegetables	-1 .66	-1 .52	-1 .37	-1 .65	-1 .50	-1 .35
Other nonenergy crops	-0.41	-0.40	-0.40	-0.42	-0.41	-0.41

[1] 2022 reference base year.

RFS-2 = Renewable Fuel Standard.

S1: tax credits retained; LP: Low oil price, HP: High oil price. S2: tax credits reduced by 50 percent; LP: Low oil price; HP: High oil price. S3: tax credits eliminated; LP: Low oil price; HP: High oil price.

Source: USAGE model simulation.

Table 6. Impact of RFS-2 on U.S. Farm Exports in 2022[1]

Commodity	S1-LP	S2-LP	S3-LP	S1-HP	S2-HP	S3-HP
	Percent change					
Corn	-3.82	-3.36	-2.85	-3.81	-3.36	-2.86
Rice	-2.16	-2.03	-1.89	-2.10	-1.98	-1.85
Cotton	-1.30	-1 .28	-1.26	-1.33	-1.31	-1.30
Wheat	-0.96	-0.95	-0.95	-0.99	-0.99	-1.00
Other feed grains	-0.54	-0.60	-0.67	-0.59	-0.65	-0.72
Soybeans	-0.79	-0.85	-0.93	-0.77	-0.84	-0.92
Other oilseeds	-1.15	-1.13	-1.12	-1.17	-1.15	-1.15
Dairy	-1.87	-1.74	-1.58	-1.97	-1.83	-1.68
Beef	-1.76	-1.65	-1.54	-1.84	-1.74	-1.63
Other livestock	-1.25	-1.18	-1.11	-1.30	-1.24	-1.17
Tobacco	-3.74	-3.44	-3.15	-3.47	-3.20	-2.92
Fruits	-2.71	-2.53	-2.35	-2.75	-2.58	-2.41
Tree nuts	-2.21	-2.07	-1.93	-2.31	-2.18	-2.04
Vegetables	-3.30	-3.09	-2.87	-3.46	-3.26	-3.05
Other crops	-1.77	-1.70	-1.62	-1.80	-1.73	-1.66

[1] 2022 reference base year.

RFS-2 = Renewable Fuel Standard.

S1: tax credits retained; LP: Low oil price, HP: High oil price. S2: tax credits reduced by 50 percent; LP: Low oil price; HP: High oil price. S3: tax credits eliminated; LP: Low oil price; HP: High oil price.

Source: USAGE model simulation.

Table 7. Impact of RFS-2 on U.S. Farm Imports in 2022[1]

Commodity	S1-LP	S2-LP	S3-LP	S1-HP	S2-HP	S3-HP
	Percent change					
Corn	12.96	11.03	8.91	12.82	10.91	8.82
Rice	4.48	4.08	3.68	4.00	3.63	3.27
Cotton	1.61	1.57	1.54	1.46	1.43	1.41
Wheat	0.85	0.86	0.87	0.71	0.72	0.74
Other feed grains	0.84	0.96	1.10	0.86	0.98	1.13
Soybeans	0.80	0.79	0.80	0.72	0.71	0.73
Other oilseeds	1.46	1.31	1.16	1.44	1.29	1.13
Dairy	1.84	1.73	1.61	2.00	1.89	1.77

Table 7. (Continued)

Commodity	S1-LP	S2-LP	S3-LP	S1-HP	S2-HP	S3-HP
	Percent change					
Beef	1.10	1.04	0.97	1.19	1.13	1.07
Other livestock	1.55	1.50	1.44	1.65	1.61	1.55
Fruits	0.80	0.78	0.75	0.87	0.85	0.82
Tree nuts	1.29	1.16	1.03	1.33	1.21	1.08
Vegetables	1.68	1.57	1.45	1.77	1.66	1.55
Other crops	1.25	1.14	1.02	1.21	1.11	1.00

[1] 2022 reference base year.

RFS-2 = Renewable Fuel Standard.

S1: tax credits retained; LP: Low oil price, HP: High oil price. S2: tax credits reduced by 50 percent; LP: Low oil price; HP: High oil price. S3: tax credits eliminated; LP: Low oil price; HP: High oil price.

Source: USAGE model simulation.

Table 8. Impact of RFS-2 on U.S. Fertilizer in 2022[1]

Item	S1-LP	S2-LP	S3-LP	S1-HP	S2-HP	S3-HP
	Percent change					
Output	6.27	5.70	5.18	6.26	5.70	5.18
Imports	10.31	9.43	8.61	10.31	9.43	8.61
Exports	4.34	3.93	3.56	4.29	3.88	3.50
Domestic price	0.01	-0.03	-0.07	-0.05	-0.09	-0.13

[1] 2022 reference base year.

RFS-2 = Renewable Fuel Standard.

S1: tax credits retained; LP: Low oil price, HP: High oil price. S2: tax credits reduced by 50 percent; LP: Low oil price; HP: High oil price. S3: tax credits eliminated; LP: Low oil price; HP: High oil price.

Source: USAGE model simulation.

LIMITATIONS AND FURTHER CONSIDERATIONS

Some limitations of this study are important in the analysis of the RFS-2. The modeling framework does not measure benefits from increased energy security or impacts associated with environmental change. Obtaining actual production data for second-generation energy crops and other advanced biofuels is still premature. How energy crops might affect regional allocation of land deserves more study. The underlying cost structure of energy crops and the differences in production costs by region could influence regional shifts in crop production. The report did not consider additional land that could potentially be used that is currently under the Conservation Reserve Program (CRP) or forest land. Furthermore, this study does not consider future developments in the rest of the world that could affect commodity supply and demand for the U.S. farm and food sector.

Another shortcoming of the analysis is in addressing public expenditures on research and development for the technological change, which would require the model to link technological change to expenditures explicitly. In addition, the study does not account for the long-term costs of infrastructure to support the biomass and biofuels industry. For example, such an accounting would involve including the costs of developing the infrastructure to support the transportation system for biomass and ethanol. Comprehensive infrastructure costs for meeting the RFS-2 are not yet known. In addition, the public costs associated with mitigating risks and financing of renewable energy investments were not taken into consideration. Capturing all future benefits and costs is beyond the scope of this study.

CONCLUSION

Increasing energy independence has been an aspiration for the United States since the early 1970s with the first major oil price spike. However, during much of the last 25 years, economic incentives for developing alternative energy has been limited because of low petroleum prices that made alternatives uncompetitive. Petroleum price increases were only temporary as oil supplies were restored within months. Low prices and policy incentives were not adequate to stimulate development of renewable fuels. The RFS-2 mandate could face similar challenges in the future.

Although global economic fluctuation creates uncertainty for near-term energy prices, technological advances play a key role in the success of a permanent and viable alternative energy industry. This analysis does not address the likelihood of meeting the RFS-2 timetable, but it attempts to provide estimates for whether and under what conditions the U.S. economy would benefit from meeting the mandate in 2022. Technological progress could enable biofuels to become competitive with petroleum, providing benefits to the U.S. economy. The larger the value of displaced petroleum for each dollar of biomass produced, the greater the benefit would accrue to the U.S. economy.

The challenge over the next decade is more likely to confront the uncertainty of short-term conditions that fall in the way of achieving long-term gains. Policies that provide incentives for producers may mitigate some risks associated with market uncertainties, whereas tax credits are more likely to reduce economic welfare. This study shows that, even with tax credits, technological progress could offset losses by raising welfare in meeting the RFS-2.

Appendix 1. Developing the Reference Scenario

The base year of the U.S. Applied General Equilibrium (USAGE) model for this study is 2005. However, the reference scenario and analysis of the Renewable Fuel Standard (RFS-2) through 2022 require that projections be made for that year (2022). The modeling framework adopts projections for the U.S. economy to develop a reference scenario. Long-term macroeconomic and energy projections for the U.S. economy are used to simulate economic growth. In our reference scenario, variables are "targeted" to match projections determined by other projection models used by the U.S. Department of Energy (DOE) and the U.S. Department of Labor. The reference scenario, also known as a "forecast scenario" in the USAGE model is based upon projected variables, such as gross domestic product (GDP), consumption, investment, population, labor force, petroleum prices, exports, and imports. These variables are exogenous variables in the reference scenario because they are determined independently from the model. The scenario serves as a neutral growth scenario where the RFS-2 is not met and is consistent with the DOE's projections. In simulating the effects of the RFS-2, most macroeconomic variables are determined by the USAGE model and reported as deviations from the reference scenario.

Based on long-term projections, U.S. economic growth, trade, and consumption spending are expected to resume growth after the 2008-10 global economic downturn. Economic growth in the United States is assumed to take place as "business as usual" through 2022. GDP is expected to grow by 2.96 percent per year and by 64.2 percent between 2005 and 2022 (app. table 1). Private consumption in the United States is expected to grow at an annual rate of 2.86 percent. The dollar is expected to continue to weaken. The weaker dollar is expected to slow import growth to 4.25 percent per year and increase export growth to 5.5 percent per year. A weakening dollar would further reduce the U.S. trade deficit.

Appendix Table 1. Reference Scenario: Key U.S. Macroeconomic Variables, 2005-22

Item	Annual growth	Accumulative growth
	Percent change	
Private consumption	2.86	61.5
Labor supply	0.91	16.7
Capital stock	2.96	64.1
Real investment	3.20	70.9
Government consumption	1.85	36.5
GDP real	2.96	64.1
Population	0.85	15.6
Exports	5.50	148.5
Imports	4.25Z	102.9
Consumer price index	2.39	49.4

Source: U.S. Department of Energy and U.S. Department of Labor.

Projections for energy variables under the reference scenario are also determined outside the USAGE model for making projections for the year 2022 (app. table 2). As vehicle fuel efficiency standards are raised, it is expected that total motor fuel consumption will fall (4.6 percent) between 2005 and 2022. Domestic production of petroleum is expected to rise over the projection period, whereas fuel imports are expected to decrease. It is assumed that, without the RFS-2, corn-based ethanol would double between 2005 and 2022, reaching 8 billion gallons. The analysis assumes that 7 billion additional gallons of corn-based ethanol are produced due to the RFS-2 mandate.

In our reference scenario, we adopt two alternative crude oil prices. We use an oil price of $80 per barrel (*Low Price*) as a low bound corresponding to ethanol's energy value for replacing gasoline and $101 (*High Price*) per barrel as an upper bound, which is directly from DOE's reference case for imported crude petroleum. Appendix table 3 shows projections from imposing the macroeconomic and the energy component of DOE's 2022 *reference* scenario on the U.S. economy under the two alternative prices for crude oil—that is, $80 and $101 per barrel. The projections are expressed as percentage changes from the base year (2005). Under the reference scenario, we assume that 8 billion gallons of conventional corn-based ethanol are available without the RFS-2, which means that corn ethanol would need to expand by 94.3 percent from a base of 4.12 billion gallons in 2005 (app. table 3). The price of gasoline would rise by 73 percent from the base year in 2005. However, the price of *blended* motor fuels (gasoline with ethanol) would rise by a lesser amount under the projection that the real price of ethanol would be $2.12 per gallon in 2022. Imported petroleum oil would fall with the rise in the price of imported petroleum, starting from $40 per barrel in 2005.

Appendix Table 2. Reference Scenario: U.S. Energy-Related Sectors, 2005-22

Item	Annual growth	Accumulative growth
	Percent change	
Natural gas consumption	0.76	13.67
Motor fuels consumption	-0.28	-4.60
Gasoline exports	0.84	15.24
Diesel exports	0.84	15.24
Crude oil exports	1.63	31.70
Gasoline imports	-2.21	-31.60
Diesel imports	-2.21	-31.60
Other petroleum imports	-2.21	-31.60
Natural gas imports	-0.69	-11.17
Electric services	1.61	31.22
Diesel production	1.45	27.70
Other petroleum fuels production	0.23	3.90
Crude petroleum production	1.13	21.10
Natural gas production	0.59	10.52
Corn ethanol production	4.16	100.00

Source: U.S. Department of Energy.

Appendix Table 3. Reference Scenario: Key U.S. Macroeconomic Variables, 2005-22

Item	Oil price $80/barrel		Oil price $101/barrel	
	Price	Quantity	Price	Quantity
	Percent change			
Gasoline	73.2	-0.2	109.0	-0.2
Motor fuels (blended gasoline)	69.3	0.1	103.5	0.1
Ethanol	0.0	94.3	0.0	94.3
Crude oil imports	100	-2.7	152.5	-2.7

[1] 2005 base year.

Source: USAGE model simulation.

APPENDIX 2. ENERGY PRICE PROJECTIONS

Future petroleum prices are likely to depend on the growth of the world economy and the underlying longrun costs for producing petroleum. The reasons for higher petroleum prices in the long run are discussed in this section.

The last time the world economy experienced a major downturn that curbed global output, trade, and energy demand was more than a quarter of a century ago. For much of the last 25 years, low prices for petroleum limited development of alternative energy. Even as world growth resumed after the 1982 recession, energy prices did not recover. Between 1985 and 2004, the real price of gasoline in the United States remained flat, rarely rising above $2 per gallon. If those same conditions were to persist over the next decade, they would weaken incentives for developing biofuels. Although the recent recession is comparable to the downturn of 1982, a repeat of the 1980s, with suppressed and prolonged stagnation in petroleum prices, is not likely because, since then, a number of factors have altered world energy markets.

The supply of petroleum has always been subject to instability from geopolitical forces. Price surges in the past were mainly from supply disruptions. High prices were rarely sustainable for more than several months at a time. However, in the recent decade, petroleum prices were elevated over a multiyear period, even without a major supply disruption. Strong demand and sluggish supply contributed to higher prices.

World energy demand continues to grow, particularly in developing countries. Their share of world GDP is now much larger than it was in the 1980s, and their energy use per dollar of GDP (energy intensity) is higher than it is in developed countries. In 1990, China and India accounted for 5 percent of demand for global liquid fuels, but their share increased to 11 percent by 2005 (app. figure 1). Economic growth was particularly strong for both of these countries between 2004 and 2008 and a key driver for raising petroleum prices during this time. By 2025, China and India are expected to account for 17 percent of global liquid fuel consumption, surpassing total consumption for all of Europe.

By 2015, other developing countries are expected to surpass the United States in liquid fuel consumption. Demand for liquid petroleum fuels is expected to remain nearly flat for Europe, Japan, and the United States between 2010 and 2025. The United States is expected to consume just 5 percent more in liquid fuels between 2007 and 2030, from 20.6 million barrels to 21.6 million barrels. Given this trend, and assuming improvement in fuel efficiency standards for vehicles, energy intensity is likely to fall in the United States. Developed countries would account for less than 50 percent of the world's liquid fuel consumption, whereas developing countries would drive nearly all of the world's increase in liquid fuel consumption. As consumers in developing countries switch to personal motorized transportation, per capita consumption of energy increases. Globally, transportation fuel use is expected to account for 74 percent of the increase in liquid fuel use. At the same time, industrial sectors are more likely to adopt efficient technologies that are lowering energy intensity.

The longrun cost of supplying petroleum fuels is more likely to rise for the following reasons. First, new investment would be needed for the world's oil producers to expand beyond current capacity. Second, the cost of extracting oil from proven oil reserves is more likely to rise. Third, oil exploration and development, which may require riskier investments that require higher rates of return, would also drive costs up.

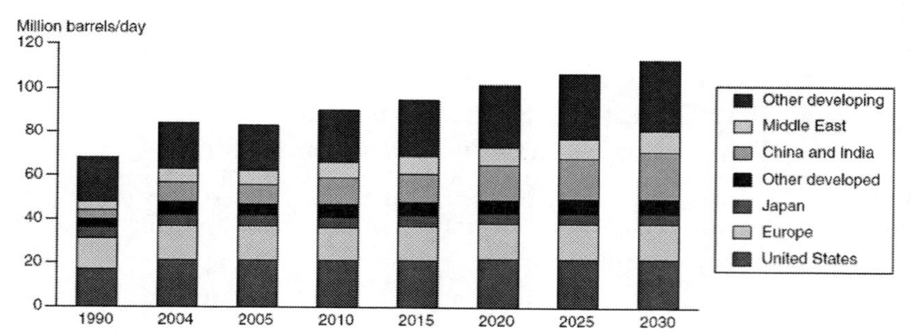

Source: U.S. Department of Energy, 2009.

Appedix Figure 1. Projected consumption of liquid fuels, 1990-2030.

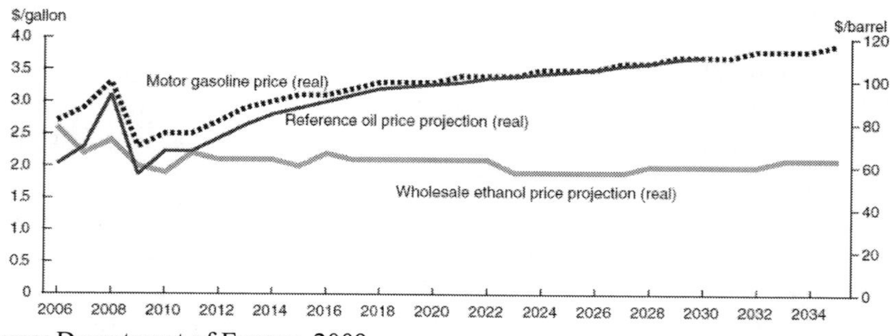

Source: Department of Energy, 2009.

Appedix Figure 2. Fuel price projections, 2006-30.

How much costs increase would depend on the source of supply because the cost structure of each petroleum source varies, depending on the geographic origin. Liquid fuels come from three basic sources: conventional petroleum from the Organization of the Petroleum Exporting Countries (OPEC), conventional petroleum from non-OPEC suppliers, and unconventional fuels, which could be supplied by both OPEC and non-OPEC sources.

Conventional petroleum supplied by OPEC is the lowest cost and most accessible liquid fuel. However, accessibility is a problem for conventional fuels supplied by non-OPEC countries. The lack of prospects for large conventional petroleum reserves in non-OPEC countries could constrain production. Development of small fields is a possibility but would come at high cost. Liquid fuels from unconventional sources are generally more expensive than those from conventional sources. Unconventional sources include oil sands (bitumen), coal-to-liquids (CTL), gas-to-liquids (GTL), and extra-heavy oil from Venezuela and Mexico.

In addition to rising costs of exploration and extraction of conventional and unconventional liquid fuels, OPEC's decision for maintaining its own market share in the world market adds to uncertainty. Even if OPEC maintains its current market share (about 40 percent) by increasing production to meet demand with other producers, the price of crude oil is more likely to continue its upward movement after 2010 (app. figure 2). The price in 2022 would be $101 per barrel in real terms (U.S. Department of Energy, 2009). If OPEC could restrain supply, the price of crude oil would be higher. Limited supply from OPEC would raise the costs of producing crude oil as it encourages production of higher cost liquid fuels by non-OPEC and unconventional sources.

In the long term, without advances in cost-reducing technology, the cost of imported crude oil will largely drive the wholesale price of gasoline (app. figure 2). Technological advances in producing alternative fuels would keep ethanol prices from rising. The longrun costs of producing alternatives fuels, like second-generation biofuels, are expected to decrease over the next decade. Although the exact timing is unknown, a growing share of ethanol is expected to be produced by second-generation biofuels as costs fall. Even as production of second-generation ethanol would expand over the decade, the real price of ethanol is not expected to increase over the long term. Improved efficiency in producing biomass and converting it to ethanol would be the key reasons for the widening price gap between ethanol and gasoline over the next two decades.

Appendix 3. Modifications for Modeling Energy Fuels

Production of cellulosic ethanol from dedicated energy crops does not yet exist on a commercial scale. This study does not attempt to make annual forecasts of production of dedicated energy crops. However, to fully implement the RFS-2 scenario, we need to make reasonable assumptions about future locations for growing dedicated energy crops. We assume that the potential for supplying dedicated energy crops varies by location because of the characteristics of the land (such as soil conditions), topography, climatic conditions, and competition with other crops.

To provide better treatment for farmland in the USAGE model, this study adopted a geographical categorizations scheme for the United States, known as Farm Resource Regions (FRR). The Farm Resource Regions were developed by USDA's Economic Research Service for depicting geographic specialization in production of U.S. farm commodities. It is used for addressing economic and resource issues in U.S. agriculture. The FRR designations consist of nine regions with boundaries defined by information on U.S. farm characteristics, county-level production, and land resource information. Regional boundaries are determined by the clustering of similar farm characteristics intersected with similar geographic climatic conditions. This type of regional grouping delineates geographical farm specialization more precisely than what would be possible using State-level groupings and aggregate farm production. For example, the Heartland region comprises three entire States but includes areas of five other States. The partial inclusion is because of the differences in farm characteristics and the underlying crop growing patterns. Each FRR has a distinct crop mix underpinning the uniqueness of each region's geography.

This study does not attempt to treat distinct types of perennial grasses in modeling cellulosic biomass. A perennial energy crop commodity, switchgrass is used to represent all related species of grasses used as dedicated energy crops in the USAGE model. Dedicated perennial energy crops may comprise multiple herbaceous species of which switchgrass is one. Although switchgrass has high potential for being grown in many regions, other species could be more suitable, depending on local conditions. Reliance on a single species, such as switchgrass, poses production risks for epidemic pest and disease outbreak. More than one species likely will be grown in any one location. Multiplicity of herbaceous species can also provide stability of yields over time. Among the promising species of perennial grasses are *Miscanthus*, reed canary grass, tall fescue, and Bermuda grass (Biomass Research and Development Board, 2008a). *Miscanthus* has higher yields than switchgrass and greater profit potential for a wide range of growing regions. The potential for dedicated energy crops varies by each FRR region (app. table 4). Regions with the highest potential are the Mississippi Portal, Eastern Uplands, the Heartland, and the Southern Seaboard. These regions would account for 80 percent of total production of dedicated energy crops. The "pre-simulated" shares reflect the potential for growing dedicated energy crops. The "post-simulated" shares reflect production based on economic returns. The USAGE model determines changes in the allocation of land for each crop based on profit maximization. Higher profits translate into higher returns to land. As the relative returns to land with greatest potential increase from competition among competing crops, land is reallocated to the most productive uses. Production can be limited as profitability diminishes from increased competition for limited land in any one region.

The USAGE model uses multi-level nests of the CRESH (constant ratio of elasticities of substitution, homothetic) functional form, representing each industry's production structure. Each pair of inputs is governed by a common parameter for all pairs of inputs. This structure, however, can pose some limitations in allowing flexibility in the substitution of intermediate inputs for ethanol. For example, cellulosic ethanol made from dedicated energy crops is highly substitutable with ethanol made from crop residues. However, other intermediate inputs, such as chemicals and enzymes, would not be substitutable with other materials from biomass. The reason for a separate cellulosic material industry is to permit a high degree of substitution of crop residue material with energy crop material but not with other intermediates, such as chemicals and enzymes. In the USAGE model, a fictional cellulosic material industry produces a product that is a combination of switchgrass and crop residue purchased by the cellulosic ethanol industry (app. figure 3). Unlike other conventional industries, the production of cellulosic material does not employ labor or capital (value-added).

Crop residues for cellulosic ethanol production could potentially come from multiple crops, such as corn, wheat, sorghum, barley, and rice. Milbrandt (2005) estimated that approximately 157 million dry tons of biomass for producing cellulosic ethanol was available in 2005. Availability of crop residues is a function of several factors, including the crop-to-residue ratio, moisture content, and alternative uses, such as animal feed and compost.[6] In this study, the supply of crop residue comes from corn only. We assume that 1 ton of corn stover, on average, is produced for every ton of corn. USDA guidelines for soil erosion require that a minimum of 30 percent be left on the ground for anti-erosion coverage. The model uses a mixed complementarity approach to capture demand and supply relationships. Supply is governed by the feed grain industry facing a joint-profit maximizing decision for corn and corn stover.

Appendix Table 4. Estimates of the Distribution of Perennial Energy Crops for Meeting the RFS-2 in 2022

Region	Pre-simulation	Post-simulation
	Percent	
Heartland	19.0	19.0
Northern Crescent	7.0	8.7
Northern Great Plains	6.0	7.4
Prairie Gateway	6.0	6.4
Eastern Uplands	20.0	19.6
Southern Seaboard	9.0	9.8
Fruitful Rim	2.0	2.4
Basin and Range	0.0	0.0
Mississippi Portal	32.0	26.6
Total	100.0	100.0

RFS-2 = Renewable Fuel Standard.
Source: USAGE model simulation.

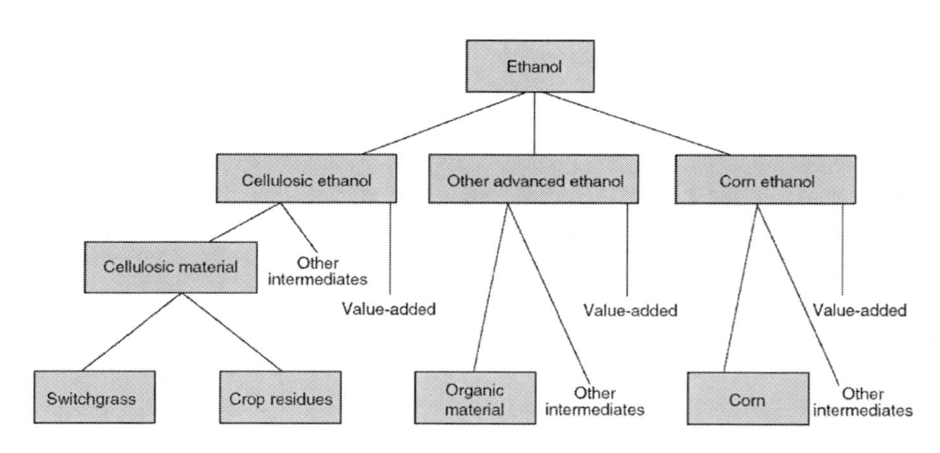

Appedix Figure 3. Ethanol production structure in USAGE model.

APPENDIX 4. SENSITIVITY ANALYSIS FOR IMPORTED CRUDE PETROLEUM

The impact of meeting the RFS-2 on the U.S. economy will depend partly on the responsiveness of the supply of crude oil in the rest of the world. The reduction in U.S. demand for petroleum reduces both quantity supplied and the price for crude oil. The more responsive the supply is for crude petroleum, the smaller the decrease in the price of crude oil. Smaller declines in the price of oil translate into lower GDP and household welfare. The import supply elasticity captures the net effect of both the demand by non-oilproducing countries and the supply of oil-producing countries. Household welfare from the impact of the RFS-2 is positive over the entire range of import supply elasticities (10-0.5) (app. table 5).

Appendix Table 5. Sensitivity Analysis of the Import Supply Elasticity for Crude Petroleum in Meeting the RFS-2

Import supply elasticity	Crude oil imports	Crude oil price	Motor fuels output	Gasoline output	Welfare	GDP
	Percent change					
10	-18.0	-2.6	2.3	-20.0	0.084	-0.026
9	-17.9	-2.8	2.3	-20.0	0.087	-0.025
8	-17.7	-3.0	2.3	-19.9	0.090	-0.024
7	-17.5	-3.2	2.4	-19.9	0.093	-0.023
6	-17.3	-3.5	2.4	-19.8	0.098	-0.021
5	-17.0	-4.0	2.5	-19.7	0.105	-0.019
4	-16.0	-5.2	2.8	-19.4	0.123	-0.013
3	-15.4	-6.1	2.9	-19.2	0.136	-0.009
2	-14.2	-7.6	3.2	-18.9	0.159	-0.001
1	-11.1	-11.3	4.0	-18.0	0.216	0.016
0.5	-7.8	-15.1	4.8	-17.0	0.277	0.035

RFS-2 = Renewable Fuel Standard.
Source: USAGE model simulation.

However, GDP becomes negative for elasticities greater than 1. This study adopts a midpoint import supply elasticity of 5. Although GDP and welfare are higher with the greater drop in petroleum, energy independence is reduced as imports of crude petroleum fall less and motor fuels output is greater.

REFERENCES

[1] Biomass Research and Development Board. (2008a). *The Economics of Biomass Feedstocks in the United States: A Review of the Literature*, Occasional Paper No. 1, October.

[2] Biomass Research and Development Board. (2008b). *Increasing Feedstock Production for Biofuels: Economic Drivers, Environmental Implications and the Role of Research*, December.

[3] de Gorter, Harry & David R. Just. (2010). "The Social Costs and Benefits of Biofuels: The Intersection of Environmental, Energy and Agricultural Policy," *Applied Economic Perspectives and Policy*, *32* (1), 4-32.

[4] De La Torre Ugarte, D. G., Walsh, M. E., Shapouri, H. & Slinsky, S. P. (2003). *The Economic Impacts of Bioenergy Crop Production on U.S. Agriculture, Agricultural Economic Report No. 816*, U.S. Department of Agriculture, Office of the Chief Economist, Office of Energy Policy and New Uses.

[5] Dixon, P. B. & Rimmer, M. T. (2002). *Dynamic General Equilibrium Modelling for Forecasting and Policy*, Amsterdam: North Holland Publishing Company.

[6] Dixon, Peter B., Stefan Osborne & Maureen T. Rimmer. (2007). "The Economy-Wide Effects in the United States of Replacing Crude Petroleum with Biomass," *Energy and Environment, 18(6)*, 709-22, November.

[7] Lubowski, R., Bucholtz, S., Claassen, R., Roberts, M., Cooper, J., Gueorguieva, A. & Johannson, R. (2006). *Environmental Effects of Agricultural Land-Use Change: The Role of Economics and Policy*, Economic Research Report No. 25, U.S. Department of Agriculture, Economic Research Service.

[8] Milbrandt, A. (2005). *A Geographic Perspective on the Current Biomass Resource Availability in the United States*, Technical Report NREL/ TP-560-39 181, National Renewable Energy Laboratory.

[9] National Academies Press. (2009). *Liquid Transportation Fuels from Coal and Biomass: Technological Status, Costs, and Environmental Impacts*, The National Academy of Sciences.
[10] Perrin, R., Vogel, K., Schmer, M. & Mitchell, R. (2008). "Farm-Scale Production Cost of Switchgrass for Biomass," *Bioenergy Research 1(1)*, 91-97.
[11] Rajagopal, D., Sexton, S., Hochman, G. & Zilberman, D. (2009). "Recent Developments in Renewable Technologies: R&D Investment in Advanced Biofuels," *Annual Review of Resource Economics, 1*, 621-44.
[12] Sheldon, I. & Roberts, M. (2008). "U.S. Comparative Advantage in Bioenergy: A Heckscher-Ohlin-Ricardian Approach," *American Journal of Agricultural Economics, 90(5)*, 1233-38.
[13] Tyner, W. E. (2007). "Policy Alternatives for the Future Biofuels Industry," *Journal of Agricultural & Food Industrial Organization 5(2)*, Article 2.
[14] Tyner, W. E. & Taheripour, F. (2007). "Future Biofuels Policy Alternatives," Paper presented at the Farm Foundation/USDA conference on Biofuels, Food, and Feed Tradeoffs, St. Louis, MO, April.
[15] U.S. Department of Energy, Energy Information Administration & Office of Energy Analysis and Forecasting. (2008). Annual Energy Outlook 2008 with Projections to 2030.
[16] U.S. Department of Energy, Energy Information Administration & Office of Energy Analysis and Forecasting. (2009). Annual Energy Outlook 2010 Early Release Overview.
[17] U.S. Government Accountability Office. (2009). Biofuels: Potential Effects and Challenges of Required Increases in Production and Use, GAO-09-446.
[18] Walsh, M. E., Perlack, R. L., Turhollow, A., De La Torre Ugarte, D., Becker, D. A., Graham, R. L., Slinsky, S. E. & Ray, D. E. (1999). Biomass Feedstock Availability in the United States: 1999 State Level Analysis, Working Paper, Oak Ridge National Laboratory.
[19] Winston, A. R. (2009). Enhancing Agricultural and Energy Sector Analysis in CGE in Modeling: An Overview of Modifications to the USAGE Model, General Paper No. G-180, Centre of Policy Studies and the Impact Project, Monash University, Australia, January, http://www.monash.edu. au/policy/ftp/workpapr/g- 1 80.pdf

End Notes

[1] The U.S. Government Accountability Office (2009) suggests that the Volumetric Ethanol Excise Tax Credit (VEETC), a 45-cent-per-gallon Federal tax credit for ethanol, may not be needed to stimulate conventional corn ethanol production because production capacity is near the RFS limit of 15 billion gallons per year.

[2] In this report, household consumption and household welfare are used interchangeably.

[3] U.S. imports of petroleum oil and petroleum-based products make up about 11 percent of total U.S. imports.

[4] Reformulated or blended fuel is produced by the "motor fuels" industry in the USAGE model.

[5] In this analysis, we assume that the consumer preference for higher blends of motor fuels would not change as a result of the RFS-2.

[6] See Biomass Research and Development Board (2008b) and De La Torre et al. (2003) for the potential supply of biomass in the United States.

In: Biofuel Use in the U.S.
Editors: S. Alonso and M. R. Ortega
ISBN: 978-1-62100-441-7

Chapter 2

BIOFUELS: CHALLENGES TO THE TRANSPORTATION, SALE, AND USE OF INTERMEDIATE ETHANOL BLENDS

United States Government Accountability Office

WHY GAO DID THIS STUDY

U.S. transportation relies largely on oil for fuel. Biofuels can be an alternative to oil and are produced from renewable sources, like corn. In 2005, Congress created the Renewable Fuel Standard (RFS), which requires transportation fuel to contain 36 billion gallons of biofuels by 2022. The most common U.S. biofuel is ethanol, typically produced from corn in the Midwest, transported by rail, and blended with gasoline as E10 (10 percent ethanol). Use of intermediate blends, such as E15 (15 percent ethanol), would increase the amount of ethanol used in transportation fuel to meet the RFS. The Environmental Protection Agency (EPA) recently allowed E15 for use with certain automobiles.

GAO was asked to examine (1) challenges, if any, to transporting additional ethanol to meet the RFS, (2) challenges, if any, to selling intermediate blends, and (3) studies on the effects of intermediate blends in automobiles and nonroad engines. GAO examined government, industry, and academic reports; interviewed Department of Energy (DOE), EPA, and other government and industry officials; and visited research centers.

WHAT GAO RECOMMENDS

GAO recommends, among other things, that EPA determine what additional research is needed on the effects of intermediate blends on UST systems. EPA agreed with the recommendation after GAO revised it to clarify EPA's planned approach.

WHAT GAO FOUND

According to government and industry officials, the nation's existing rail, truck, and barge infrastructure should be able to transport an additional 2.4 billion gallons of ethanol to wholesale markets by 2015—enough to meet RFS requirements. Later in the decade, however, a number of challenges and costs are projected for transporting additional volumes of ethanol to wholesale markets to meet peak RFS requirements. According to EPA estimates, if an additional 9.4 billion gallons of ethanol are consumed domestically by 2022, several billion dollars would be needed to upgrade rail, truck, and barge infrastructure to transport ethanol to wholesale markets.

GAO identified three key challenges to the retail sale of intermediate blends:

- *Compatibility*. Federally sponsored research indicates that intermediate blends may degrade or damage some materials used in existing underground storage tank (UST) systems and dispensing equipment, potentially causing leaks. However, important gaps exist in current research efforts—none of the planned or ongoing studies on UST systems will test actual components and equipment, such as valves and tanks. While EPA officials have stated that additional research will be needed to more fully understand the effects of intermediate blends on UST systems, no such research is currently planned.
- *Cost*. Due to concerns over compatibility, new storage and dispensing equipment may be needed to sell intermediate blends at retail outlets. The cost of installing a single-tank UST system compatible with intermediate blends is more than $100,000. In addition, the cost of installing a single compatible fuel dispenser is over $20,000.

- *Liability*. Since EPA has only allowed E15 for use in model year 2001 and newer automobiles, many fuel retailers are concerned about potential liability issues if consumers misfuel their older automobiles or nonroad engines with E15. Among other things, EPA has issued a proposed rule on labeling to mitigate misfueling.

DOE, EPA, and a nonfederal organization have provided about $51 million in funding for ten studies on the effects of intermediate blends on automobiles and nonroad engines—such as weed trimmers, generators, marine engines, and snowmobiles—including effects on performance, emissions, and durability. Of these studies, five will not be completed until later in 2011. Results from a completed study indicate that such blends reduce a vehicle's fuel economy (i.e., fewer miles per gallon) and may cause older automobiles to experience higher emissions of some pollutants and higher catalyst temperatures. Results from another completed study indicate that such blends may cause some nonroad engines to run at higher temperatures and experience unintentional clutch engagement, which could pose safety hazards.

ABBREVIATIONS

CRC	Coordinating Research Council, Inc.
DOE	Department of Energy
DOT	Department of Transportation
E10	fuel blend containing approximately 10 percent ethanol
E15	fuel blend containing approximately 15 percent ethanol
E20	fuel blend containing approximately 20 percent ethanol
E85	fuel blend containing 70 percent to 83 percent ethanol
EISA	Energy Independence and Security Act
EPA	Environmental Protection Agency
NIST	National Institute of Standards and Technology
NREL	National Renewable Energy Laboratory
ORNL	Oak Ridge National Laboratory
OSHA	Occupational Safety and Health Administration
RFS	Renewable Fuel Standard

UL	Underwriters Laboratories
USDA	Department of Agriculture
UST	underground storage tank

June 3, 2011

The Honorable Fred Upton
Chairman
The Honorable Joe Barton
Chairman Emeritus
Committee on Energy and Commerce
House of Representatives

The Honorable Cliff Stearns
Chairman
Subcommittee on Oversight and Investigations
Committee on Energy and Commerce
House of Representatives

The Honorable Michael C. Burgess
The Honorable Greg Walden
House of Representatives

The U.S. transportation sector is almost entirely dependent on petroleum products refined from crude oil—primarily gasoline and diesel fuels. In 2009, this sector consumed the equivalent of about 14 million barrels of oil per day, or over 70 percent of total U.S. consumption of petroleum products. To meet the demand for crude oil and petroleum products, the nation imported, on a net basis, about 52 percent of the petroleum products consumed in 2009.[1]

Biofuels can be an alternative to petroleum-based transportation fuels and are produced from renewable sources, primarily corn, sugar cane, and soybeans. The United States is the world's largest producer of biofuels. The federal government has promoted the domestic production and use of biofuels through tax incentives since the 1970s and, more recently, through a Renewable Fuel Standard (RFS). The Energy Policy Act of 2005, which created the RFS, generally required transportation fuels in the United States to contain renewable fuels, such as ethanol and biodiesel.[2] The Energy Independence and Security Act (EISA) of 2007 expanded the RFS by requiring that U.S. transportation fuels contain 9 billion gallons of renewable fuels in 2008, with renewable fuels increasing annually to 36 billion gallons in 2022.[3] The Environmental Protection Agency (EPA) is responsible for administering the RFS.

Ethanol is the most commonly produced biofuel in the United States. In 2010, the nation produced 13.2 billion gallons of ethanol, the vast majority of which came from corn. Most U.S. corn is grown in the Midwest, and ethanol is generally produced in relatively small biorefineries near corn-producing areas. Unlike petroleum products, which are primarily transported to wholesale terminals by pipelines, ethanol is transported to wholesale terminals by a combination of rail, tanker truck, and barge. At the terminals, most ethanol is currently blended as an additive in gasoline to make fuel blends containing up to 10 percent ethanol (called E10). Finally, the blended fuel is transported via tanker truck to retail fueling outlets.

In a 2009 report, we identified fuel-blending limits as a challenge to expanded ethanol consumption.[4] We stated that the nation may soon reach a "blend wall"—the upper limit to the total amount of ethanol that can be blended into U.S. gasoline, given current constraints. At the time, the blend wall existed partly because under EPA's implementation of the Clean Air Act, fuels containing more than 10 percent ethanol were prohibited from being introduced for use with the vast majority of U.S. automobiles.[5] This created a blend wall at approximately 10 percent of total U.S. fuel consumption. If the volume of renewable fuels required by the RFS increased above this 10 percent threshold, the fuel industry would not be able to meet the RFS using only E10. We noted that one option to address the blend wall is to use "intermediate" ethanol blends such as E15 or E20 (generally 15 percent or 20 percent ethanol).[6]

In March 2009, a group of ethanol manufacturers petitioned EPA to allow E15 into commerce. The Clean Air Act prohibits the introduction of fuels that are not substantially similar to gasoline. However, the Act authorizes EPA to grant a waiver of this prohibition for a fuel if it does not cause vehicles or engines to exceed emission standards over their useful life. EPA issued two decisions on E15. The first, issued in October 2010, allowed E15 for use in model year 2007 and newer automobiles. The second, issued in January 2011, allowed E15 for use in model years 2001 through 2006 automobiles. EPA did not allow E15 for use in older automobiles or nonroad engines (such as lawn mowers, chainsaws, and boats), motorcycles, or heavy-duty gasoline engines. EPA cited insufficient test data to support the use of E15 in these engines, as well as engineering concerns that older vehicles and nonroad engines may not maintain compliance with emission standards if operated on E15.[7]

In light of the potential use of intermediate ethanol blends, you asked us to review their potential effects. Our objectives were to (1) determine the challenges, if any, associated with transporting additional volumes of ethanol to wholesale markets to meet RFS requirements; (2) determine the challenges, if any, associated with selling intermediate ethanol blends at the retail level; and (3) examine research by federal agencies into the effects of intermediate ethanol blends on the nation's automobiles and nonroad engines.

To determine the challenges associated with transporting additional volumes of ethanol to wholesale markets to meet RFS requirements, we reviewed relevant literature and reports from federal government agencies—including EPA, the Department of Energy (DOE), and the Department of Transportation (DOT)—industry associations, and academic organizations and interviewed their relevant officials and representatives. To determine the challenges associated with selling intermediate ethanol blends at the retail level, we reviewed relevant literature and reports from federal and state government agencies—including EPA, DOE, the Department of Labor's Occupational Safety and Health Administration (OSHA), and the California Air Resources Board—government laboratories, and industry associations and interviewed their relevant officials and representatives. To examine research by federal agencies into the effects of intermediate ethanol blends on the nation's automobiles and nonroad engines, we reviewed relevant reports and studies from government and private laboratories, including DOE's National Renewable Energy Laboratory (NREL) in Colorado and Oak Ridge National Laboratory (ORNL) in Tennessee and interviewed their relevant officials. We also conducted site visits to NREL, ORNL, and a private laboratory to discuss testing results. Due to ongoing litigation over EPA's decision to allow E15 for use in certain automobiles, we did not make any determination in this report of the adequacy of federal testing efforts for automobiles. In addition, we interviewed officials from EPA, DOE, and representatives from relevant industry associations. A more detailed description of our scope and methodology is presented in appendix I.

We conducted this performance audit between April 2010 and June 2011 in accordance with generally accepted government auditing standards. Those standards require that we plan and perform the audit to obtain sufficient, appropriate evidence to provide a reasonable basis for our findings and conclusions based on our audit objectives. We believe that the evidence obtained provides a reasonable basis for our findings and conclusions based on our audit objectives.

BACKGROUND

The RFS, as defined by EISA, distinguishes between ethanol derived from corn starch (known as corn ethanol) and advanced biofuels—defined as a renewable fuel other than corn ethanol that meets certain criteria. For example, to qualify as an advanced biofuel, a biofuel must reduce lifecycle greenhouse gas emissions by at least 50 percent compared to the gasoline or diesel fuel it displaces.[8] According to the RFS, most advanced biofuels must be produced from cellulosic materials, which can include perennial grasses, crop residue, and the branches and leaves of trees. In addition, some advanced biofuels must be produced from biomass-based diesel, which generally includes any diesel made from biomass feedstocks, such as soybeans. As shown in figure 1, the volume of corn ethanol included under the RFS is capped at 15 billion gallons by 2015 and is fixed thereafter

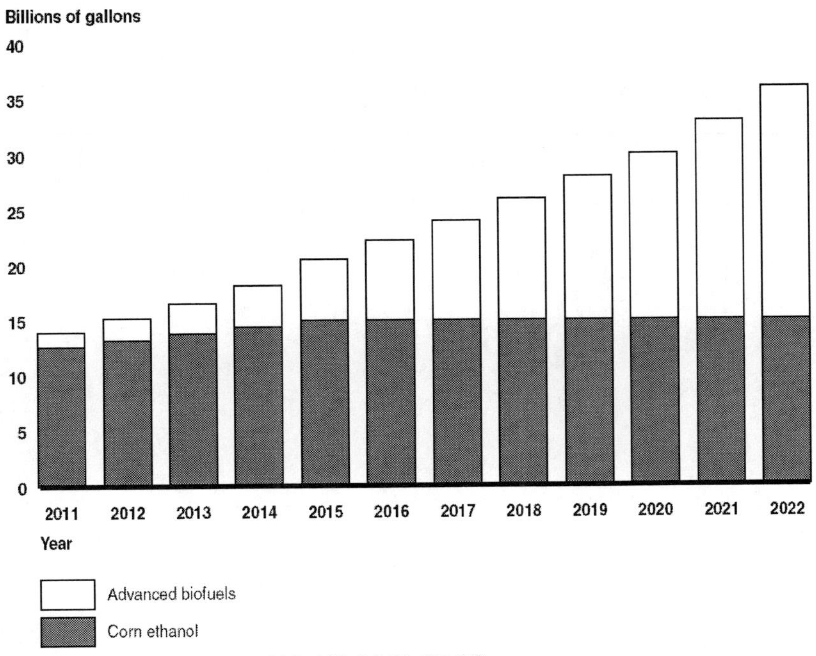

Source: EISA, Pub. Law No. 110-140 § 202 (2007).

Figure 1. Volume Requirements Established by the Renewable Fuel Standard under the Energy Independence and Security Act.

Pursuant to this provision, EPA has already lowered the RFS requirements for cellulosic biofuel, from 250 million gallons to 6.6 million gallons for 2011, mostly due to the small number of companies with the potential to produce cellulosic biofuel on a commercial scale.[9]

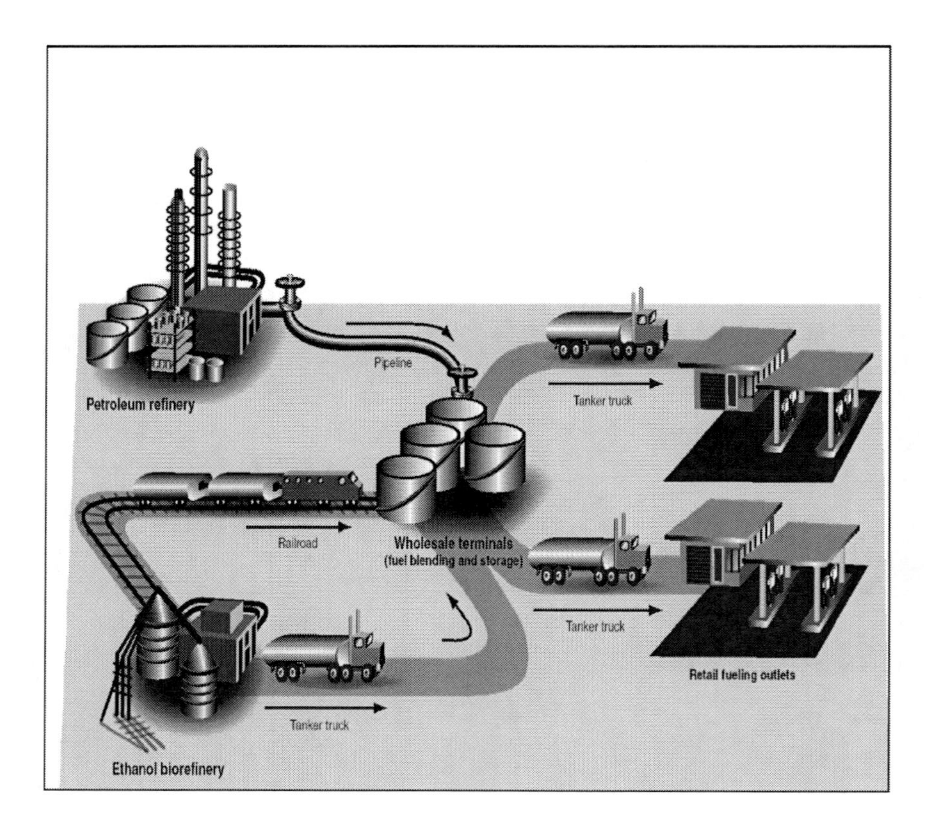

Source: GAO.

Note: Other means of transportation are also used to move petroleum and ethanol products to wholesale terminals. For example, for ethanol, barges are also used to a limited extent.

Figure 2. Primary Transportation of Petroleum Products and Ethanol from Refineries to Retail Fueling Outlets.

As shown in figure 2, the infrastructure used to transport petroleum fuels from refineries to wholesale terminals in the United States is different from that used to transport ethanol. Petroleum-based fuel is primarily transported from refineries to terminals by pipeline.[10] In contrast, ethanol is transported to terminals via a combination of rail cars, tanker trucks, and barges.[11] According to DOE estimates, there are approximately 1,050 terminals in the United States that handle gasoline and other petroleum products. At the terminals, most ethanol is currently blended as an additive in gasoline to make E10 fuel blends.

However, the volume of advanced biofuels continues to grow to a total of 21 billion gallons by 2022. By comparison, the U.S. transportation sector consumed about 14 million barrels of oil per day in 2009, which translates to more than 99 billion gallons of gasoline consumed for the entire year.

The RFS generally requires that U.S. transportation fuels in 2022 contain 36 billion gallons of biofuels. In addition, at least 16 billion of the 36 billion gallons of biofuels must be cellulosic biofuels—including ethanol and diesel derived from cellulosic materials. However, under EISA, EPA is required to determine the projected available volume of cellulosic biofuel production for the year, and if that number is less than the volume specified in the statute, EPA must lower the standard accordingly.

A relatively small volume is also blended into a blend of between 70 percent to 83 percent ethanol (E85) and the remainder gasoline. E85 has a more limited market, primarily in the upper Midwest, and can only be used in flexible-fuel vehicles, which are vehicles that have been manufactured or modified to accept it.[12] After blending, the fuel is moved to retail fueling locations in tanker trucks.

There are approximately 159,000 retail fueling outlets in the United States, according to 2010 industry data.[13] This total included more than 115,000 convenience stores, which sold the vast majority of all the fuel purchased in the United States, according to industry estimates; a number of large retailers that sell fuel, such as Walmart, Costco, and several grocery chains; and some very low-volume retailers, such as marinas. In terms of ownership, single-store businesses—that is, businesses that own a single retail outlet—account for about 56 percent of the convenience stores selling fuel in the United States.

There are three primary supply arrangements between fuel retailers and their suppliers:

- *Major oil owned and operated.* About 1 percent (or 1,175) of convenience stores selling fuel in the United States are owned and operated by four major integrated oil companies—ExxonMobil, Chevron, BP, and Shell.[14]

- *Branded independent retailer.* About 52 percent of retail fueling outlets are operated by independent business owners who sell fuel under the brand of one of the major oil companies or refineries (such as CITGO, Sunoco, or Marathon).[15] These retailers sign a supply and marketing contract with their supplier to sell fuel under the brand of that supplier.
- *Unbranded independent retailer.* The remaining retail fueling outlets (about 48 percent) are operated by independent business owners who do not sell gasoline under a brand owned or controlled by a refining company. These retailers purchase gasoline from the unbranded wholesale market, which is made up of gallons not dedicated to fulfill a refiner's contracts with branded retailers.

Federal safety and environmental regulations govern the dispensing and storage of fuel at retail fueling locations. First, OSHA requires that equipment used to dispense gasoline—including hoses, nozzles, and other related aboveground components, shown in figure 3—be certified for safety by a nationally recognized testing laboratory.[16] According to OSHA officials, OSHA recognizes 17 laboratories, although Underwriters Laboratories (UL) is the main one that currently certifies equipment sold for dispensing gasoline.[17] In addition, under the Solid Waste Disposal Act, EPA requires that underground storage tank (UST) systems—including storage tanks, piping, pumps, and other related underground components, shown in figure 3—must be compatible with the substance stored in them to protect groundwater from releases from these systems. Historically, UL certification has been the primary method for determining the compatibility of USTs with EPA requirements.[18] EPA also requires fuel retailers to install equipment to detect leaks from UST systems. In total, EPA regulates approximately 600,000 active USTs at about 215,000 sites in the United States.

As shown in figure 2, the infrastructure used to transport petroleum fuels from refineries to wholesale terminals in the United States is different from that used to transport ethanol. Petroleum-based fuel is primarily transported from refineries to terminals by pipeline.[10] In contrast, ethanol is transported to terminals via a combination of rail cars, tanker trucks, and barges.[11] According to DOE estimates, there are approximately 1,050 terminals in the United States that handle gasoline and other petroleum products. At the terminals, most ethanol is currently blended as an additive in gasoline to make E10 fuel blends.

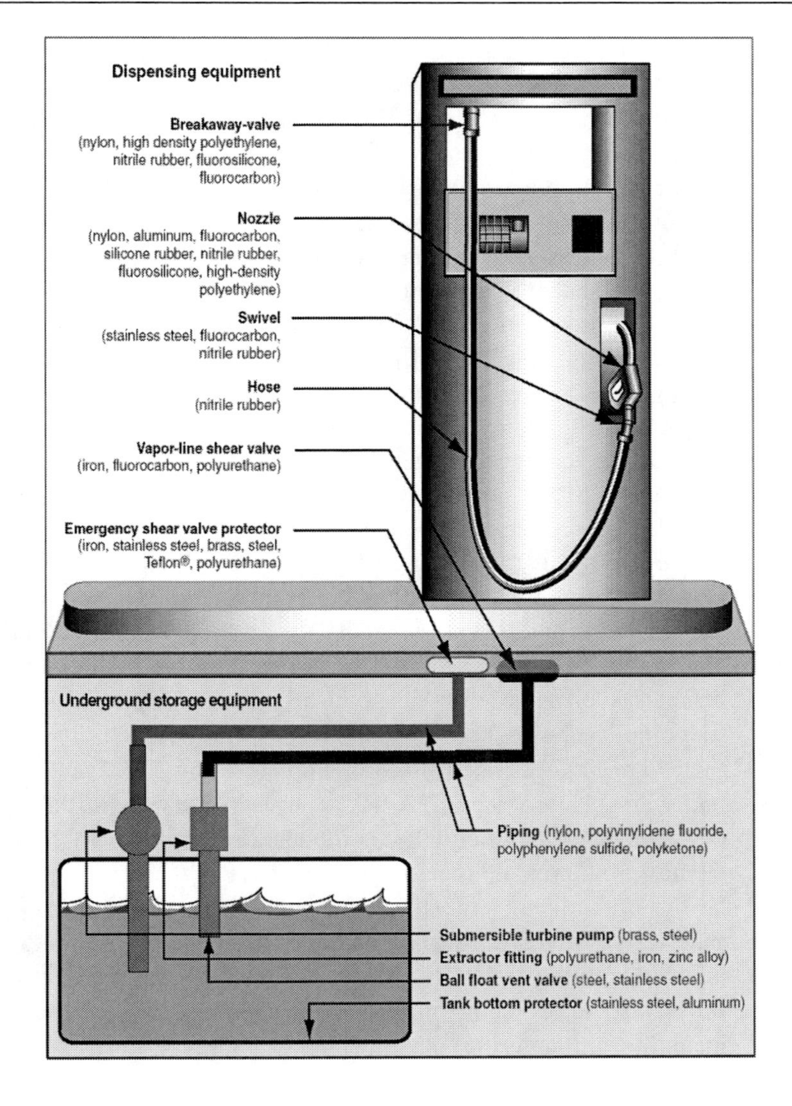

Source: GAO analysis of DOE and EPA information.

Figure 3. Examples of Typical Components and Materials in Retail Dispensing and Underground Storage Equipment.

State and local governments also play a role in regulating the safety of dispensing equipment and in implementing EPA's requirements for USTs. For example:

- The Occupational Safety and Health Act allows states to develop and operate their own job safety and health programs. OSHA approves and monitors state programs and plans, which must adopt and enforce standards that are at least as effective as comparable federal standards. According to OSHA officials, there are currently 21 states with approved plans covering the private sector that enforce health and safety standards over the dispensing of gasoline within their respective states. Four additional states operate approved state plans that are limited in coverage to the public sector.
- Various state and local fire-safety codes—which aim to protect against fires—also govern the dispensing of fuel at retail fueling outlets. While state fire marshals or state legislatures are usually responsible for developing the fire code for their respective states, some states allow local municipalities to develop their own fire codes. Fire codes normally reference or incorporate standards developed by recognized standards-development organizations, such as the National Fire Protection Association and the International Code Council.[19] State, county, and local fire marshals are responsible for enforcing the applicable fire code within their respective jurisdictions. Local officials, such as fire marshals, typically inspect dispensing equipment for compliance with both state and local fire codes.
- States are largely responsible for implementing EPA's requirements under its UST program. EPA has approved 36 states, plus the District of Columbia and Puerto Rico, to operate programs in lieu of the federal program. The remaining states have agreements with EPA to be the primary implementing agency for their programs. Typically, states rely on UL certification as the primary method for determining the compatibility of UST systems with EPA requirements. Some states also allow compatibility to be demonstrated in other ways, including through the manufacturer's approval or a professional engineering certification.

Consumers in the United States use retail fueling locations to fuel hundreds of millions of automobiles and nonroad products with gasoline engines. According to DOT data, Americans owned or operated almost 256 million automobiles, trucks, and other highway vehicles in 2008, while about 91 percent of all households owned at least 1 automobile the same year, according to U.S. Census data. Americans also owned and operated over 400 million products with nonroad engines in 2009, according to one industry association estimate. According to EPA documentation, nonroad engines are typically more basic in their engine design and control than engines and emissions control systems used in automobiles, and commonly have carbureted fuel systems[20] and air cooling, whereby extra fuel is used in combustion to help control combustion and exhaust temperatures. According to representatives from industry associations for nonroad engines, most of the small nonroad engines manufactured today rely on older technologies and designs to keep retail costs low, and all of the small nonroad engines currently being produced are designed to perform successfully on fuel blends up to E10. According to industry representatives, while it is possible to design small nonroad engines to run on a broad range of fuels, such designs would not be cost effective and could add hundreds of dollars to the price.

Challenges to Transporting Additional Volumes of Ethanol to Wholesale Markets May Require Large Investments in Infrastructure over the Next Decade

Existing ethanol infrastructure should be sufficient to transport the nation's ethanol production through 2015, according to DOT officials and industry representatives, but large investments in transportation infrastructure may be needed to meet 2022 projected consumption, according to EPA documentation. One option for doing so may be to construct a dedicated ethanol pipeline, but this option presents significant challenges.

Investments in Transportation Infrastructure May Be Needed to Meet 2022 Ethanol Consumption Levels

According to knowledgeable DOT officials and industry representatives we met with, the existing rail, truck, and barge transportation infrastructure for shipping corn ethanol to wholesale markets should be sufficient through 2015, when the volume of corn ethanol in the RFS is effectively capped at 15 billion gallons annually. This volume represents roughly a 2.4 billion gallon increase from 2011 RFS consumption targets for corn ethanol. Specifically, for rail, which transports about 66 percent of corn ethanol to wholesale markets, several DOT officials and representatives from the Association of American Railroads told us that the addition of a few billion gallons of ethanol over the near term is not expected to have a significant impact. Railroads hauled more than 220,000 rail carloads of ethanol in 2008 (the most recent year for which data are available)—which was about 0.7 percent of all the rail carloads and about 1 percent of the total rail tonnage transported that year in the United States, according to data from the Association of American Railroads. Similarly, knowledgeable DOT officials and industry representatives said there is sufficient capacity in the short term to transport additional volumes of corn ethanol via trucks, which transport about 29 percent of corn ethanol to wholesale markets, and barges, which transport roughly 5 percent, to meet RFS requirements.

In contrast, the existing infrastructure may not be sufficient to handle the ethanol production that is projected after 2015. The RFS generally requires transportation fuels in the United States to contain 21 billion gallons of advanced biofuels, including a large quantity of cellulosic ethanol, by 2022. In a 2010 regulatory impact analysis, EPA assessed the impacts of an increase in the production, distribution, and use of ethanol and other biofuels sufficient to meet this requirement.[21] In its assessment, EPA used three scenarios or "control cases" to project the amounts and types of renewable fuels to be produced domestically and imported from 2010 through 2022.[22] Under its "primary" control case, EPA projected that by 2022, the United States would produce and import over 22 billion gallons of ethanol, comprising 15 billion gallons of domestically produced corn ethanol, almost 5 billion gallons of domestically produced cellulosic ethanol, and over 2 billion gallons of imported ethanol.[23] EPA also estimated the number of facilities that would need to be built or modified, as well as the number of additional vehicles that would need to be purchased. Under its primary control case, EPA estimated that the necessary spending on transportation infrastructure due to increased ethanol consumption would be approximately $2.6 billion. According to EPA's analysis:

- *For rail.* EPA estimated that approximately $1.2 billion would be needed for an additional 8,450 rail tanker cars ($760 million) and the construction of new train facilities ($446 million). EPA projected that biofuels transport will constitute approximately 0.4 percent of the total tonnage for all commodities transported by the freight rail system through 2022. Sixteen percent of the nation's freight rail system would be affected by biofuels shipments, and that portion (mostly along rail corridors radiating out of the Midwest) would see a 2.5 percent increase in traffic.

- *For trucks.* EPA estimated that approximately $87 million would be needed for an additional 480 tank trucks.
- *For barges.* EPA estimated that approximately $198 million would be needed for an additional 32 barges ($45 million), and the configuration of barge facilities (a projected $153 million). EPA stated that it does not anticipate a substantial fraction of biofuels will be transported via barge over the inland waterway system. In addition, the agency projected that a total of 30 ports will receive significant quantities of imported ethanol from Brazil and Caribbean Basin Initiative countries by 2022.
- *For wholesale terminals.* EPA estimated that $1.15 billion in investments would be needed, primarily to modify vapor recovery equipment (at a cost of $1 million for each terminal that does not already handle ethanol). Other modifications would include the installation of new storage tanks, modification of existing tanks, and modification of tank-truck unloading facilities.

EPA stated that the United States will face significant challenges in accommodating the projected increases in biofuels production by 2022, but it concluded that the task would be achievable at the wholesale level. For example, the agency stated that it believed overall freight-rail capacity would not be a limiting factor to the successful implementation of RFS requirements.

However, while this task may be achievable, it is likely to be increasingly difficult because of congestion on U.S. transportation networks. We and others have reported that congestion is constraining the capacity and increasing the costs of U.S. rail and highway transportation. For example, we reported in 2008 that neither rail nor highway capacity had kept pace with recent increases in demand, leading to increased costs.[24] We also cited a study by the Association of American Railroads, which predicted that without system improvements, the expected increases in rail volume by 2035 will cause 30 percent of primary rail corridors to operate above capacity and another 15 percent at capacity. The study stated the resulting congestion might affect the entire country and could shut down the national rail network. In addition, we noted that many of the highways used heavily by trucks to move freight are already congested, and congestion is expected to become a regular occurrence on many intercity highways. Finally, we noted that ports are likely to experience greater congestion in the future as more and larger ships compete for limited berths.

One Option for Transporting Additional Volumes of Ethanol—Constructing a Dedicated Pipeline—Presents Significant Challenges

If overall ethanol production increases enough to fully meet the RFS over the long term, one option to transport it to wholesale markets would be through a dedicated ethanol pipeline. Over many decades, the United States has established very efficient networks of pipelines that move large volumes of petroleum-based fuels from production or import centers on the Gulf Coast and in the Northeast to distribution terminals along the coasts. However, the existing networks of petroleum pipelines are not well suited for the transport of billions of gallons of ethanol. Specifically, as shown in figure 4, ethanol is generally produced in the Midwest and needs to be shipped to the coasts, flowing roughly in the opposite direction of petroleum-based fuels. The location of renewable fuel production plants (such as biorefineries) is often dictated by the need to be close to the source of the raw materials and not by proximity to centers of fuel demand or existing petroleum pipelines.

Existing petroleum pipelines can be used to ship ethanol in some areas of the country. For example, in December 2008, the U.S. pipeline operator Kinder Morgan began transporting commercial batches of ethanol along with gasoline shipments in its 110-mile Central Florida Pipeline from Tampa to Orlando. Pipeline owners would face the same technical challenges and costs that Kinder Morgan representatives reported facing, including the following:[25]

- *Compatibility.* Ethanol can dissolve dirt, rust, or hydrocarbon residues in a petroleum pipeline and degrade the quality of the fuel being shipped. It can also damage critical nonmetallic components, including gaskets and seals, which can cause leaks. In order for existing pipelines to transport ethanol, pipeline operators would need to chemically remove residues and replace any components that are not compatible with ethanol. According to DOT officials, the results from two research projects sponsored by that agency have identified specific actions that must be taken on a wide variety of nonmetallic components commonly utilized by the pipeline industry.[26]

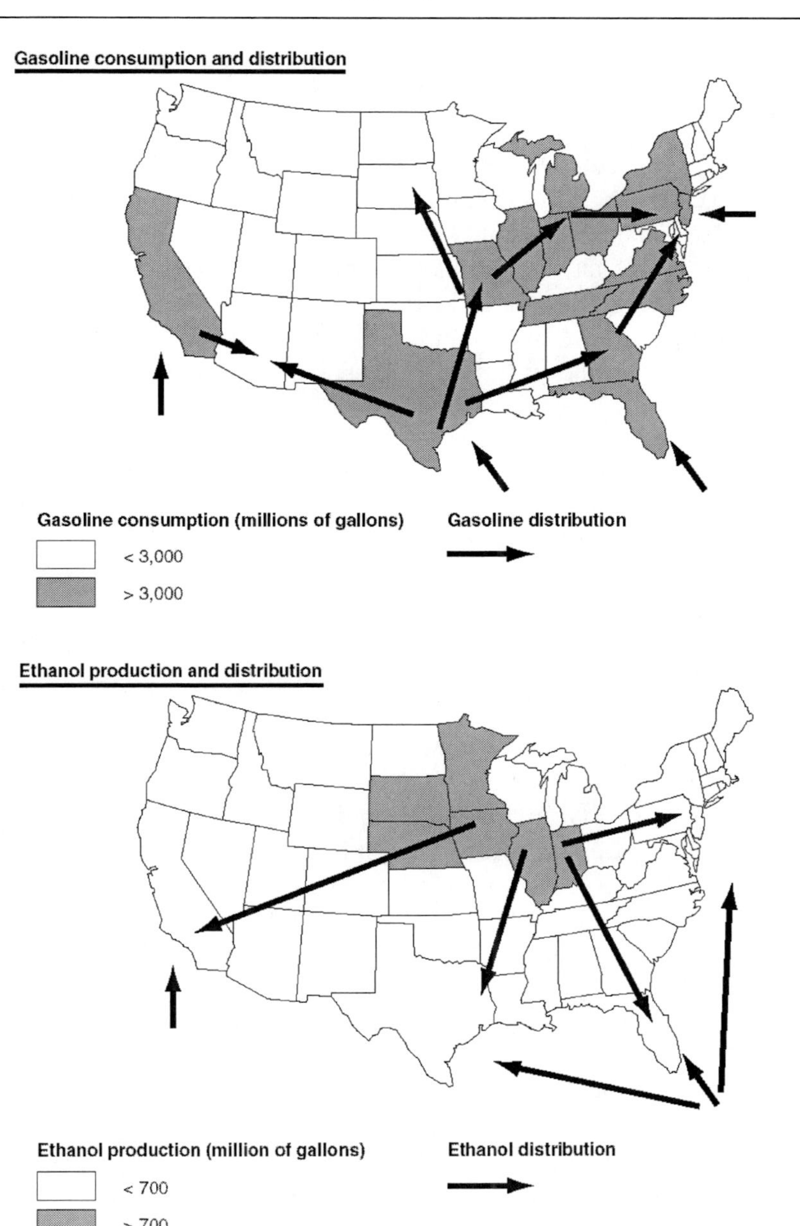

Source: DOE, *Report to Congress: Dedicated Ethanol Pipeline Feasibility Study* (Washington, D.C., March 2010).

Figure 4. Distribution Patterns for Gasoline and Ethanol.

- *Stress corrosion cracking.* Tensile stress and a corrosive environment can combine to crack steel. The presence of ethanol increases the likelihood of this in petroleum pipelines. Over the past 2 decades, approximately 24 failures due to stress corrosion cracking have occurred in ethanol tanks and in production-facility piping having steel grades similar to those of petroleum pipelines. According to DOT officials, the results from nine research projects sponsored by that agency have targeted these challenges and produced guidelines and procedures to prevent or mitigate stress corrosion cracking. As a result, pipelines can safely transport ethanol after implementing the identified measures, according to DOT officials.[27]
- *Attraction of water.* Ethanol attracts water. If even small amounts of water mix with gasoline-ethanol blends, the resulting mixture cannot be used as a fuel or easily separated into its constituents. The only options are additional refining or disposal.

Some groups have proposed the construction of a new pipeline dedicated to the transportation of ethanol. For example, in February 2008, Magellan Midstream Partners, L.P. (Magellan) and Buckeye Partners, L.P. (Buckeye) proposed building a new pipeline from the Midwest to the East Coast.[28] According to this proposal, the pipeline would gather ethanol from three segments: (1) Iowa, Nebraska, and South Dakota; (2) Illinois, Michigan, and Minnesota; and (3) Indiana and Ohio. Ethanol would be transported to demand centers in New England, the Mid-Atlantic, Virginia, and West Virginia.

The federal government has studied the feasibility of building a pipeline similar to the one proposed by Magellan. Specifically, under section 243 of EISA, DOE (in collaboration with DOT) issued a study in March 2010 that examined the feasibility of constructing an ethanol pipeline linking large East Coast demand centers with refineries in the Midwest.[29] The report identified a number of significant challenges to building a dedicated ethanol pipeline, including the following:

- *Construction costs.* Using recent trends in and generally accepted industry estimates for pipeline construction costs, DOE estimated that an ethanol pipeline from the Midwest to the East Coast could cost about $4.5 million per mile. While DOE assumed that the construction of 1,700 miles of pipeline would cost more than $3 billion, it did not model total project costs beyond $4.25 billion in the report.

- *Higher transportation rates.* Based on the assumed demand for ethanol in the East Coast service area and the estimated cost of construction, DOE estimated the ethanol pipeline would need to charge an average tariff of 28 cents per gallon, substantially more than the current average rate of 19 cents per gallon, for transporting ethanol using rail, barge, and truck along the same transportation corridor.
- *Lack of eminent domain authority.* DOE estimated that siting a new ethanol pipeline of any significant length will likely require federal eminent domain authority, which currently does not exist for ethanol pipelines.

  DOE's report concluded that a dedicated ethanol pipeline can become a feasible option if there is (1) adequate demand for the ethanol (approximately 4.1 billion gallons per year for the hypothetical pipeline assessed) and (2) government financial incentives to help defray the large construction costs.

Challenges Due to Regulations, Technical Issues, and Cost Could Slow the Retail Sale of Intermediate Ethanol Blends

We identified several challenges to selling intermediate ethanol blends at the retail level. First, federal and state regulations governing health and environmental concerns must be met before these blends are allowed into commerce, and fuel-testing requirements to meet these regulations may take 1 year or more to complete. Second, according to knowledgeable federal officials and UL representatives, federal safety standards do not allow ethanol blends over E10 to be dispensed at most retail fueling locations, and federally sponsored research has indicated potential problems with the compatibility of intermediate ethanol blends with existing dispensing equipment. Third, according to EPA and several industry representatives, the compatibility of many UST systems with these fuels is uncertain, and retailers will need to replace any components that are not compatible if they choose to store intermediate blends. Fourth, industry associations representing various groups, such as fuel retailers and refiners, are concerned that, in selling intermediate ethanol blends, fuel retailers may face significant costs and risks, such as upgrading or replacing equipment.

Federal and State Regulations Need to Be Met Prior to the Introduction of Intermediate Blends

According to knowledgeable EPA officials within the Office of Transportation and Air Quality, the regulatory process for allowing an intermediate ethanol blend into commerce could take 1 year or more. As described in table 1, the Clean Air Act, among other things, establishes a comprehensive regulatory program aimed at reducing harmful emissions from on- and off-road vehicles and engines and the fuels that power them. According to EPA officials, this regulatory program would apply to the introduction of new fuels, including E15 and other intermediate blends.

Although intermediate ethanol blends higher than E15 would need to meet all of these requirements, E15 has already partly met the first two. EPA partially granted a fuel waiver allowing E15 for use in model year 2001 and newer automobiles, and EPA officials told us the agency has no plans to revise its regulations for certifying detergents for E15 because it currently has not determined any detergent-related issues different from E10. According to EPA officials, the remaining two requirements have not yet been completed for E15 but are in the process of being addressed, specifically:

- Health-effects testing similar to that performed for E10 could take 2 years or more to register intermediate ethanol blends, depending on variables such as the availability of testing laboratories. According to EPA officials, EPA received information on February 18, 2011, from an ethanol industry representative contending that the health-effects testing previously performed for E10 is an adequate substitute for E15. According to recent Congressional testimony, EPA expects to finish reviewing the information by the middle of 2011.[30]
- EPA would have to update the regulations for its reformulated gasoline program, which do not currently allow fuel manufacturers to certify batches of gasoline containing greater than 10 percent ethanol by volume. In November 2010, EPA proposed a rule that would, among other things, update the model to allow for reformulated gasoline containing up to 15 percent ethanol by volume.[31] According to EPA officials, EPA expects to issue a final rule sometime in 2011.

Table 1. Federal Fuel Requirements that Affect the Introduction of New Fuels

Type of requirement	Description
Fuel waiver	Under the Clean Air Act, the introduction into commerce of new fuels and fuel additives that are not substantially similar to the fuels and fuel additives specified by EPA regulations for testing the compliance of vehicles and engines with EPA emission standards is prohibited. However, EPA may waive this prohibition if a demonstration is made that the fuel or fuel additive will not cause or contribute to vehicles and engines failing to meet emission standards over their useful lives.
Detergent certification	EPA regulations implementing the Clean Air Act require fuel manufacturers to certify any detergent added to gasoline to prevent the accumulation of deposits in engines and fuel systems. Fuel manufacturers must use EPA-approved test fuels to certify the effectiveness of new detergents. EPA regulations currently require these test fuels to contain 10 percent ethanol by volume.
Fuel registration and health-effects testing	The Clean Air Act and EPA regulations require fuel manufacturers and importers to register new fuels and fuel additives prior to introducing them into commerce. Registration involves providing a chemical description of the product and certain technical, marketing, emissions, and health-effects information, which EPA uses to identify likely combustion and evaporative emissions that may pose concerns about health risk. However, EPA regulations allow registrants to submit evidence that prior health-effects testing is reasonably comparable to the results expected for a new fuel or fuel additive.
Reformulated gasoline certification	The Clean Air Act requires reformulated gasoline to be sold in areas of the country with the worst smog pollution, which include large areas of California and the Northeast. Reformulated gasoline must meet specific EPA emission standards. Fuel manufacturers must use an EPA-approved model to certify that new reformulated fuels meet applicable standards.

Source: GAO analysis of EPA information.

In addition to federal regulations, many states have established regulations or statutes related to transportation fuels, according to a 2010 industry report.[32] In particular, many state regulations or statutes contain references to specific industry standards for fuel published by a recognized standards development organization, including ASTM International and the National Institute of Standards and Technology (NIST), according to the report and knowledgeable NIST officials we interviewed. These standards, however, are only relevant to E10, and neither organization has published any standards related to the use of intermediate ethanol blends up to E85. Therefore, before allowing intermediate ethanol blends into commerce, the states that reference existing ASTM International or NIST standards would have to either (1) enact new statutes or regulations that no longer reference the existing standards or (2) wait for ASTM International or NIST to update their standards related to intermediate ethanol blends. Either option could take more than a year to implement, according to knowledgeable officials from NIST and the California Air Resources Board.

OSHA Regulations Prohibit Using Most Existing Dispensing Equipment with Intermediate Blends, and Research Suggests Compatibility Issues

In general, federal safety standards do not allow ethanol blends over E10 to be dispensed with existing equipment at most retail fueling locations. Specifically, OSHA requires that all equipment used to dispense gasoline be certified for safety by a nationally recognized testing laboratory. UL, the only such laboratory that has developed standards for certifying dispensing equipment, did not publish safety standards specifically for intermediate ethanol blends until August 2009,[33] and no UL-certified dispensing equipment was available for use with these blends until 2010.[34] Dispensing equipment manufactured earlier has been certified for blends up to E10, and UL does not recertify equipment that has already been certified to an existing UL standard, according to several UL representatives. Moreover, UL does not retroactively certify manufactured or installed equipment to new safety standards because it cannot monitor whether the equipment has been modified by, for example, aging or maintenance. As a result, according to knowledgeable OSHA officials and several UL representatives, the vast majority of existing retail dispensers in the United States are not approved for use with intermediate ethanol blends under OSHA's safety regulations.

Until recently, UL and OSHA were each exploring ways to allow fuel retailers to use existing dispensing equipment with intermediate ethanol blends while still meeting OSHA's safety regulations. For example, in a February 2009 announcement, UL stated that existing dispensing equipment—certified for use with E10—could be used with blends containing up to 15 percent ethanol, based on data the company had collected. According to the announcement, UL did not find any significant incremental risk of damage to existing equipment between E10 and fuels with a maximum of 15 percent ethanol. In addition, several OSHA officials told us in November 2010 that the agency was at the early stages of evaluating several options—such as implementing a grace period on planned enforcement activities or developing an enhanced inspection and maintenance program for a limited time—that would allow existing dispensing equipment to be approved for use with E15.

However, results from federally sponsored research indicate potential problems with the use of intermediate ethanol blends with some existing dispensing equipment. A DOE-commissioned report prepared by UL was issued in November 2010 on the compatibility of intermediate blends with new and used dispensing equipment certified for blends up to E10.[35] According to the report, although various components generally performed well with the testing fluid, some of the components tested (including valve assemblies and nozzles) demonstrated a reduced level of safety, performance, or both when exposed to the testing fluid. This was mostly due to the failure of certain nonmetal components, such as gaskets and seals. In March 2011, DOE's ORNL published a report stating that, although metal samples experienced very little corrosion, all elastomer samples (such as fluorocarbon, nitrile rubber, and polyurethane) exhibited some level of swelling and the potential to leak when exposed to testing fluids.[36]

This research has led UL and OSHA to reconsider support for the use of existing dispensing equipment with intermediate ethanol blends. In a December 2010 announcement based on this research, UL stated that it advised against the use of intermediate ethanol blends with dispensing equipment certified for E10 and, instead, recommended the use of new equipment designed and certified for use with intermediate ethanol blends. The announcement stated that UL was particularly concerned that blends over E10 could lead to the degradation of gaskets, seals, and hoses and could cause leaks. In addition, several OSHA officials told us that, as a result of this research, the agency is re-evaluating its plan to explore ways to allow fuel retailers, under certain conditions, to use existing dispensing equipment with intermediate blends.

However, OSHA's position on this issue remains unclear, and it is uncertain when the agency will establish a definitive position. On the one hand, according to several OSHA officials we talked with, the vast majority of existing retail dispensers in the United States are not approved for use with intermediate ethanol blends under OSHA's safety regulations. On the other hand, these officials also stated that OSHA is still developing its position on the use of existing dispensing equipment with intermediate blends. While these officials said that strict enforcement of current OSHA requirements for dispensing equipment seems more like an option now, they did not provide any time frames for when OSHA would finalize its position, nor how it planned on communicating a decision to fuel retailers and other interested parties.

The Compatibility of Many UST Systems with Intermediate Blends Is Unclear

According to our discussions with knowledgeable federal officials and several industry association representatives, the compatibility of many existing UST systems with intermediate ethanol blends is unclear for two main reasons—many fuel retailers have older equipment and lack records, and recent federally sponsored research indicates potential problems with the use of intermediate blends. Retail fueling outlets generally have two or more UST systems, according to industry association representatives, and each system contains a large number of components and materials. According to EPA documentation and knowledgeable EPA officials within the Office of Underground Storage Tanks, many existing USTs range in age from 1 to 40 years and contain components certified to a range of UL standards, which typically have evolved over time, or have been approved by the manufacturer for varying uses.[37] Because these systems are buried underground, visually inspecting some components for compatibility is impossible without excavating them. Thus, fuel retailers, along with state and federal inspectors, primarily rely on recordkeeping to verify UST system compatibility with the fuel stored in them.

However, inadequate recordkeeping may make it difficult for retailers with older stations to verify UST system compatibility with intermediate ethanol blends. For example, according to EPA documentation, knowledgeable EPA officials, and a representative from the Society of Independent Gasoline Marketers of America, many fuel retailers do not have complete records of all their UST equipment, particularly those with stations having several previous owners. Furthermore, many installation companies and component manufacturers may have gone out of business, according to EPA documentation, which could make verification particularly challenging. Recognizing this issue, EPA announced in November 2010 that it plans to issue guidance that would clarify its compatibility requirements for UST systems storing ethanol blends higher than 10 percent.[38] In its announcement, EPA also solicited public feedback on the extent of the challenges fuel retailers face in demonstrating existing UST systems' compatibility with intermediate ethanol blends and on alternatives that would sufficiently protect human health and the environment. EPA officials said the agency expects to issue guidance sometime in 2011.

Determining compatibility may be important because ongoing federal research indicates potential problems with the use of intermediate ethanol blends with some UST components. For example, according to a recent DOE report and additional results from DOE research, certain elastomers, rubbers, and other materials used in UST systems may degrade or swell excessively when exposed to intermediate ethanol blends, becoming ineffective as gaskets or seals.[39] DOE testing also indicates that a pipe-thread sealant commonly used in UST piping in the past is not compatible with any ethanol blends, which raises concerns that these components may leak when exposed to ethanol—even in lower blends, such as E10. According to the report, DOE expects to conclude this research in the near future. In addition, DOE officials said they do not expect to conduct additional research on UST components or equipment.

However, important gaps exist in current federal research efforts in this area. For example, several officials within EPA's Office of Underground Storage Tanks told us that DOE's research efforts to date have focused only on testing materials (e.g., elastomers and rubbers) and not actual components and equipment (e.g., valves and tanks) found in UST systems. In addition, according to EPA officials, while the agency plans to study the compatibility of E15 with UST systems, this research will be based on interviews with experts and not on actual testing of materials, components, or equipment. Moreover, EPA officials characterized this research effort as more of a "modeling" or scoping effort to determine the extent of any potential problems. EPA officials stated that the ability to determine the compatibility of legacy equipment with intermediate blends is limited. Nevertheless, they acknowledged that additional research will be necessary to facilitate a transition to storing intermediate ethanol blends in UST systems, including the suitability of specific UST components with intermediate blends. EPA officials told us that they are working with industry officials and federal partners to understand the impact of intermediate blends in UST systems. However, to date EPA has not developed a plan to undertake such research.

It is also unclear whether leak-detection equipment will properly detect leaks of intermediate ethanol blends. According to knowledgeable EPA officials and UL representatives, UL has not developed performance standards for leak-detection equipment used in UST systems. EPA officials explained that, while some leak-detection equipment has been approved by the manufacturer for the compatibility of its materials with intermediate ethanol blends, EPA is not certain whether the ethanol content of the fuel, in general, would affect the operability of this equipment. To address this potential problem, EPA is sponsoring research, in collaboration with manufacturers and other stakeholders, to determine which of these devices works properly with ethanol. EPA officials currently expect test results to be available by the end of 2011.

Retailers May Face Significant Costs and Risks in Selling Intermediate Blends

According to several industry associations representing various groups, such as fuel retailers and refiners, many fuel retailers may face significant costs and risks in selling intermediate ethanol blends. According to these industry representatives, retailers make very little money selling fuel—for example, the national average profit from selling gasoline last year was 9 cents per gallon, according to industry data. Most retailers make most of their profit selling merchandise such as food, beverages, and tobacco products, according to these industry representatives, and gasoline is sold below cost in some markets to attract customers to buy more profitable goods. As a result, according to several industry representatives, most retailers do not upgrade their fuel-storage and -dispensing equipment without a significant market opportunity.

For these fuel retailers, the prospect of selling intermediate ethanol blends presents several potential challenges. The first is cost. Some fuel retailers may have to spend hundreds of thousands of dollars to upgrade their equipment to store and dispense intermediate ethanol blends, for the following reasons:

- Under current OSHA regulations, most fuel retailers will need to replace at least one dispenser system to sell intermediate ethanol blends. According to estimates from EPA and several industry associations, installing a new dispenser system compatible with intermediate ethanol blends will cost over $20,000.[40] According to some industry association representatives, a typical fuel retailer has four dispensers and, therefore, would face costs exceeding $80,000 to upgrade an entire retail facility.
- Fuel retailers with inadequate records of their UST systems may have to upgrade certain UST components to demonstrate compatibility with intermediate ethanol blends. According to some industry association representatives and information from DOE's NREL, upgrading some components would be less expensive than installing an entirely new UST system. Taking this into consideration, EPA estimated an average cost of $25,000 per retail facility to make the needed changes to underground storage components.[41] However, EPA cautioned that this cost scenario is very speculative, given that the costs of modifying underground components could vary greatly. According to EPA officials, most tank owners will be able to demonstrate compatibility by replacing certain portions of the UST system that are easily accessible (such as submersible pumps, tank probes, pipe dope, and overfill valves). The costs for these upgrades, including labor, can be as low as a few thousand dollars but may increase if more extensive upgrades are required.
- According to EPA and industry estimates, the total cost of installing a new single-tank UST system compatible with intermediate ethanol blends is more than $100,000. In addition to the high costs, some industry association representatives stated that fuel retailers who have recently installed new UST systems may be particularly reluctant to replace them, especially since UST warranties can last for several decades, and the useful life of these systems can be even longer. In Florida, for example, fuel retailers were required to replace or upgrade all single-wall USTs by December 31, 2009.

A second potential challenge consists of financial and logistical limitations on the types of fuel a retailer may be able to sell. According to representatives from several industry associations, most retail fueling locations have only two UST systems, and many fuel retailers cannot install additional UST systems due to space constraints, permitting obstacles, or cost.[42]

Currently, fuel retailers with two UST systems can sell three grades of gasoline: regular, midgrade, and premium. To accomplish this, they typically use one of their tanks to store regular gasoline and the other for premium, both of which are preblended with up to 10 percent ethanol. They then use their dispensing equipment to blend fuel from both tanks into midgrade gasoline. If fuel retailers with two UST systems want to sell intermediate ethanol blends, however, they may face certain limitations. For example, fuel retailers with two UST systems who want to sell regular, midgrade, and premium gasoline could use the tanks to store regular and premium grades of an intermediate blend, such as E15. However, since EPA has only allowed E15 for use in model year 2001 and newer automobiles, these retailers would not be able to sell fuel to consumers for use in older automobiles and nonroad engines.

A third potential challenge relates to legal uncertainty among industry groups, who are concerned they could be held liable for selling intermediate ethanol blends. For example, according to representatives we interviewed from several industry associations, fuel retailers have received conflicting or confusing messages from different authorities as to whether existing dispensing equipment can be lawfully used with intermediate ethanol blends. According to these industry representatives, this confusion is partly the result of UL's 2009 announcement supporting the use of blends containing up to 15 percent ethanol with existing dispensing equipment.[43] However, even if state or local officials—such as fire marshals—approve the use of intermediate blends with existing dispensers, the retailers selling these blends would still be effectively ignoring OSHA's regulations, which require the use of equipment that has been certified for safety by a nationally recognized testing laboratory, such as UL. As a result, several industry representatives raised concerns that fuel retailers could expose themselves to lawsuits for negligence and invalidate important business agreements that may reference these safety requirements, such as tank insurance policies, state tank-fund policies, and business loan agreements.

In addition, according to representatives from several industry associations we interviewed, many fuel retailers are concerned that consumer misfueling—using intermediate ethanol blends in nonapproved engines—could raise liability issues, especially if the misfueling is associated with negative outcomes, such as diminished engine performance and safety problems. Because EPA has only allowed E15 for use in model year 2001 and newer automobiles, representatives from several industry associations stated that consumers may not be aware of the distinction between approved and nonapproved engines, or they may be confused about which fuel to use, thus complicating their experience at retail fueling outlets and increasing opportunities for misfueling. According to some industry and state government representatives, since many automobile manufacturer warranties do not cover the use of intermediate ethanol blends, even for the model year vehicles approved by EPA for E15, consumers could be held responsible for the cost of any repairs attributed to the use of E15.

One proposed method of mitigating the potential for misfueling is to label fuels at retail outlets. In November 2010, EPA issued proposed labeling requirements for ethanol blends as high as E15.[44] According to its proposed requirements, EPA is coordinating with the Federal Trade Commission, which in March 2010 proposed labeling requirements for ethanol blends containing greater than 10 percent and less than 70 percent ethanol by volume.[45] However, representatives from several industry associations have raised concerns that labeling will not adequately address potential misfueling. For example, some industry association representatives stated that some consumers will not understand the label, or the label might get lost among the other labels commonly found on dispensers. Furthermore, industry association representatives said some consumers will intentionally misfuel their automobiles if intermediate ethanol blends are cheaper. For example, industry association representatives stated some of their members have witnessed consumers using E85 in nonflex-fuel vehicles, presumably because E85 is cheaper than E10.

Federally Sponsored Studies Are Evaluating Effects of Using Intermediate Ethanol Blends in Automobiles and Nonroad Engines

With the possibility of introducing intermediate ethanol blends in the nation's motor-fuel supply, DOE began to study the effects of these fuels in automobiles and nonroad engines in 2007. Specifically, in March 2007, DOE's Office of Energy Efficiency and Renewable Energy convened a workshop of experts to evaluate progress and develop a strategy for meeting the Bush Administration's "20 in 10" initiative. The goal of the initiative was to reduce U.S. gasoline usage by 20 percent over the next 10 years through increased use of alternative fuels and improved fuel economy. One conclusion from the workshop was that increasing the ethanol content in motor fuel to E15 or E20 would be the most effective strategy over the short term.

However, based on a review of existing research, DOE's ORNL found that almost no data existed on the effects of E15 on automobiles, while only limited data existed on the effects of E20.[46]

To address this data gap, DOE began working with EPA, the Coordinating Research Council, Inc. (CRC), and other groups in 2007 to develop a list of research projects to test the effects of E15 and E20 on automobiles and nonroad engines.[47] DOE, EPA, and CRC have provided about $51 million in funding (for fiscal years 2007 through 2010) for ten research projects (see table 2).[48]

Of the six federally sponsored projects on automobiles, four projects are ongoing and are expected to be completed in 2011. Two projects have been completed—Project V1, which looked primarily at the effects of E15 and E20 on tailpipe emissions from automobiles, and Project V3, which looked primarily at the effects of E20 on evaporative emissions from automobiles. According to published reports, project findings included the following:

Table 2. Status of DOE- and EPA-Sponsored Research on Effects of Intermediate Ethanol Blends in Automobiles and Nonroad Engines

DOE project number[a]	Description	Status (as of Mar. 1, 2011)
Research on automobiles		
V1	Short-term "quick-look" emissions study	Completed. Published reports: • NREL, ORNL, *Effects of Intermediate Ethanol Blends on Legacy Vehicles and Small Non-Road Engines,* Report 1 –Updated (Golden, Colo., February 2009). • Keith Knoll, et al., "Effects of Mid-Level Ethanol Blends on Conventional Vehicle Emissions" (paper presented at SAE 2009 Powertrains Fuels and Lubricants Meeting, San Antonio, Tex., November 2009).
V2	Detailed exhaust emissions study	Ongoing. Expected date for completion of testing is May 2011. Expected date for issuing a report is October 2011.

V3	Evaporative emissions study	Completed. Published reports: • Harold Haskew & Associates, Inc., *Evaporative Emissions from In-Use Vehicles: Test Fleet Expansion* (CRC E-77-2b) Final Report EPA-420-R-10-025, a technical report prepared for the EPA, October 2010. • CRC, *Study to Determine Evaporative Emission Breakdown, Including zermeation Effects and Diurnal Emissions, Using E20 Fuels on Aging Enhanced Evaporative Emissions Certified Vehicles* (Alpharetta, Ga., December 2010).
V4	Full-life emissions study	Ongoing. Testing completed in December 2010. Expected date for issuing a report is summer 2011.
V5	Drivability study	Ongoing. Expected date for completion of testing is March 2011. Expected date for issuing a report is August 2011.
V6	Fuel-system materials compatibility study	Ongoing. Expected date for issuing a report is summer 2011.
Research on nonroad engines		
SE1	"Quick-look" emissions and temperature study	Completed. Published report: NREL, ORNL, *Effects of Intermediate Ethanol Blends.*
SE2	Full useful-life emissions and durability study	Completed. Published report: NREL, ORNL, *Effects of Intermediate Ethanol Blends.*
SE3	Chainsaw safety study	Canceled.
SE4	Marine and snow-mobile durability, emissions, and drivability study	Ongoing. Expected date for completion of testing is March 2011 (for marine engines) and August 2011 (for snowmobiles). Expected date for issuing a report is October 2011.

Source: GAO analysis of DOE, EPA, and CRC information.

[a] CRC project numbers associated with these efforts include E-77 (for V3), E-87 (for V4), E-89 (for V2), CM-138 (for V5), and AVFL-15 (for V6).

- *Exhaust emissions.* According to the 2009 DOE report for Project V1, regulated tailpipe emissions from 16 automobiles (including model years ranging from 1999 to 2007) remained largely unaffected by the ethanol content of the fuel.[49] Increasing the ethanol content of the fuel, however, resulted in increased emission of ethanol and acetaldehyde. DOE has also released all of the testing data from Project V4, which is looking at emissions testing and aging on 82 automobiles (including model years ranging from 2000 to 2009). EPA based its decision to allow E15 for use in certain automobiles partly on these results. According to EPA's decision, model year 2000 and older automobiles do not have the sophisticated emissions control systems of more recently manufactured automobiles, and there is an engineering may experience emissions increases if operated on E15.
- *Fuel economy.* According to DOE's report for Project V1, ethanol has about 67 percent of the energy density of gasoline on a volumetric basis. As a result, automobiles running on intermediate ethanol blends exhibited a loss in fuel economy commensurate with the energy density of the fuel. Specifically, when compared to using gasoline containing no ethanol, the average reduction in fuel economy was 3.7 percent using E10, 5.3 percent using E15, and 7.7 percent using E20.
- *Catalyst temperatures.* According to the 2009 report for Project V1, 9 of the 16 automobiles adjusted their air-to-fuel ratio at full power to compensate for the increased oxygen content in the ethanol-blended fuel. In these cases, the catalyst temperatures at equivalent operating conditions were lower or unchanged with ethanol. Seven of the 16 tested automobiles failed to adequately adjust their air-to-fuel ratio for the increase in oxygen with E20 fuel compared with 100 percent gasoline at full power. As a result, catalyst temperatures for these automobiles at full power were between 29°C and 35°C higher with E20 relative to gasoline. According to the report, the long-term effect of this temperature increase on catalyst durability is unknown and requires further study.
- *Evaporative emissions.* According to its 2010 report for Project V3, CRC found that intermediate ethanol blends may increase evaporative permeation emissions—fuel-related emissions that do not come from the tailpipe—in older automobiles. CRC's report was not based on statistically significant comparisons, but it noted certain trends—for example, compared to pure gasoline, E10 and E20 were associated with increased evaporative emissions.

Of the four federally sponsored projects on nonroad engines, one (SE4) is ongoing, and one (SE3) has been canceled. According to DOE, the objective of Project SE4 is to determine the effects of E15 on the safety, performance, and emissions of several popular marine and snowmobile engines. The objective of Project SE3 was to assess the effects of intermediate ethanol blends, including E15, on the safety and performance of handheld small nonroad engines, including chainsaws. However, according to DOE officials, the department decided in the summer of 2010 to defer Project SE3 indefinitely because the Outdoor Power Equipment Institute—an industry association representing small nonroad engine manufacturers and DOE's major partner on the project—declined to submit a proposal for conducting the testing. According to one official with the Institute, this decision was based, in part, on EPA's indication that it would not allow E15 for use in small nonroad engines.

The two federally sponsored projects on nonroad engines that have been completed—SE1 and SE2—were not conclusive, but indicated potential problems with the use of intermediate ethanol blends in small nonroad engines. Project SE1 was a pilot study of six commercial and residential small nonroad engines, and Project SE2 tested 22 engines over their full useful lives. According to the 2009 DOE report, the projects found that with increasing levels of ethanol:[50]

- For all engines tested, exhaust and engine temperatures generally increased.
- Three handheld trimmers had higher idle speeds and experienced unintentional clutch engagement, which DOE laboratory officials identified as a potential safety concern that can be mitigated in some engines by adjusting the carburetor.
- For all engines tested, emissions of nitrogen oxides increased and emissions of carbon monoxide decreased, while emissions of hydrocarbons decreased in most engines, but increased for some.

EPA cited results from Projects SE1 and SE2 in its decision to not allow the use of E15 in nonroad engines and other equipment. Specifically, in its October 2010 decision, EPA stated that the results of these projects indicated reasons for concern with the use of E15 in nonroad engines, particularly regarding long-term exhaust and evaporative emissions durability and materials compatibility. Moreover, the agency stated that the application for use of E15 did not provide information to broadly assess the nonroad engine and vehicle sector. EPA concluded that since there are important differences in design between the various types of nonroad engines, and since the agency was not aware of other information that would allow it to fully assess the potential impacts of E15 on the emission performance of nonroad products, it could not allow the use of E15 in these engines.

Due to ongoing litigation, we did not evaluate the adequacy of these federally sponsored projects. In November 2010, several trade groups representing the oil and gas sector and the food and livestock industries filed a lawsuit with the U.S. Court of Appeals for the District of Columbia Circuit challenging EPA's E15 waiver decision. According to the plaintiffs' statement filed in January 2011, one key issue in the lawsuit is whether EPA acted arbitrarily, capriciously, and in excess of its statutory authority by relying on data that do not provide adequate support for its conclusions, while ignoring extensive data contradicting its position. In addition, in December 2010, several trade groups representing automobile and small-engine manufacturers filed another lawsuit with the U.S. Court of Appeals for the District of Columbia Circuit challenging EPA's E15 waiver decision. The initial court documents did not provide details on these groups' rationale for challenging EPA's waiver decision.

In addition to these federally sponsored projects, some nonfederal organizations are conducting research on the effects of intermediate ethanol blends in automobiles. Appendix II provides a description of these organizations and a list of some of their published research. We did not evaluate the results of these studies.

CONCLUSION

The RFS calls for increasing amounts of biofuels to be blended in the nation's transportation fuel supply, including up to 15 billion gallons of ethanol made from corn starch and potentially billions of gallons of additional ethanol made from cellulosic sources. EPA is responsible for establishing and implementing regulations to ensure that the nation's transportation fuel supply contains the volumes of biofuels required by the RFS. The agency is also tasked with ensuring that new fuels do not cause or contribute to noncompliance with existing emissions standards when used in automobiles and nonroad products. EPA recently allowed an intermediate ethanol blend, E15, for use in model year 2001 and newer automobiles, after determining that it would not cause these automobiles to be out of compliance with emissions standards.

EPA, along with OSHA, is also responsible for ensuring that fuels are compatible and safe for use with infrastructure at fueling locations. However, the effects of intermediate ethanol blends on key components of the nation's retail fueling infrastructure—such as gaskets and seals in dispensing equipment and UST systems—are not fully understood. A recently published DOE report found that materials commonly used in gaskets and seals can swell when exposed to certain intermediate ethanol blends, potentially causing leaks.

In the case of fuel-dispensing equipment, some newer equipment meets OSHA safety regulations for use with intermediate ethanol blends, as this equipment has been tested and certified by UL for compatibility. Most existing equipment at retail fueling locations in the United States, however, is not approved for use with intermediate blends. Until recently, OSHA had been exploring ways to allow fuel retailers to use existing equipment with intermediate blends while still meeting OSHA's safety requirements. In light of the recent DOE-sponsored research, OSHA officials are re-evaluating the use of existing equipment with intermediate blends. However, the agency has not clarified when it will make an official decision. Without clarification from OSHA on how its safety regulations on fuel-dispensing equipment should be applied to fuel retailers selling intermediate ethanol blends, the retail fuel industry faces uncertainty in how it can provide such blends to consumers while meeting OSHA safety regulations.

In the case of UST systems, fuel retailers can purchase new equipment—certified by UL or the equipment manufacturer for use with intermediate ethanol blends—to meet EPA regulations for compatibility. However, many existing UST systems may not be fully compatible with intermediate blends, and inadequate records may make it difficult for many retailers to verify the compatibility of their UST systems. Due to these concerns, and in light of the recent DOE-sponsored research, EPA is in the process of issuing guidance to clarify how its UST regulations apply to the use of intermediate blends. While DOE is conducting studies on the compatibility of UST materials with intermediate blends, and while EPA plans to conduct a study limited to experts' views on the subject, EPA officials have acknowledged that additional research, including research on the suitability of specific UST components with intermediate blends, will be needed to facilitate a transition to storing intermediate ethanol blends. Without this effort, the retail fuel industry faces uncertainty in how it can provide intermediate blends to consumers.

Recommendations for Executive Action

We are making the following two recommendations:

- To reduce uncertainty about the applicability of federal safety regulations, we recommend that the Secretary of Labor direct the Assistant Secretary for Occupational Safety and Health to issue guidance clarifying how OSHA's safety regulations on fuel-dispensing equipment should be applied to fuel retailers selling intermediate ethanol blends.
- To reduce uncertainty about the potential environmental impacts of storing intermediate ethanol blends at retail fueling locations, we recommend that the Administrator of EPA determine what additional research, such as research on the suitability of specific UST components, is necessary to facilitate a transition to intermediate ethanol blends, and work with other federal agencies to develop a plan to undertake such research.

Appendix I. Scope and Methodology

To determine the challenges associated with transporting additional volumes of ethanol to wholesale markets to meet Renewable Fuel Standard (RFS) requirements, we interviewed relevant government, industry, academic, and research officials. We also reviewed relevant government reports and studies, industry reports, and academic and research literature. In particular, we asked a nonprobability sample of knowledgeable stakeholders, among other things, to discuss the challenges, if any, associated with transporting additional volumes of ethanol to wholesale markets. We also asked these stakeholders to identify key studies and other knowledgeable stakeholders on this topic. We selected these stakeholders using a "snowball sampling" technique, whereby each stakeholder we interviewed identified additional stakeholders and stakeholder organizations for us to contact. Specifically, based, in part, on our recent work, we first interviewed stakeholders from the Environmental Protection Agency (EPA); the Departments of Agriculture (USDA), Energy (DOE), and Transportation (DOT); the Renewable Fuels Association; the American Petroleum Institute; the Alliance of Automobile

Manufacturers; the Association of Oil Pipe Lines; and the Outdoor Power Equipment Institute.[51] We then used feedback from these interviews to identify additional stakeholders to interview.[52] Over the course of our work, we interviewed officials from the following federal agencies: DOE Office of the Biomass Program, DOE Office of Vehicle Technologies Program, DOT Research and Innovative Technology Administration, DOT Pipeline and Hazardous Materials Safety Administration, DOT Federal Railroad Administration, DOT Federal Motor Carrier Safety Administration, DOT Maritime Administration, EPA Office of Research and Development, EPA Office of Solid Waste and Emergency Response, EPA Office of Transportation and Air Quality, USDA Agricultural Research Service, and USDA Economic Research Service. We also interviewed state officials from the Minnesota State Fire Marshal Division and the Office of North Carolina State Fire Marshal. We interviewed industry representatives from the following organizations: the American Petroleum Institute, the Association of American Railroads, the Association of Oil Pipe Lines, Growth Energy, Independent Fuel Terminal Operators Association, Kinder Morgan, the National Petrochemical and Refiners Association, the National Tank Truck Carriers, American Trucking Associations, and the Renewable Fuels Association. We also made several attempts to speak with representatives from an industry association representing barge operators but were not able to schedule an interview during the time frame of our audit. Finally, we interviewed academic and research stakeholders from Carnegie Mellon University, the Energy Policy Research Foundation, the James A. Baker III Institute for Public Policy of Rice University, the Pipeline Research Council International, and TRC Energy Services. During these interviews, knowledgeable stakeholders identified a number of studies related to our work. Of these studies, we identified the following three studies as being directly relevant to our scope of analysis: (1) the National Commission on Energy Policy's Task Force on Biofuels Infrastructure, (2) EPA's Renewable Fuel Standard Program (RFS2) Regulatory Impact Analysis, and (3) DOE's Report to Congress: Dedicated Ethanol Pipeline Feasibility Study.[53] We examined these three studies and determined that they are sufficiently reliable for our purposes based on interviews with contributors to these studies, comparisons of estimates with other sources, and checking selected calculations.

To determine the challenges associated with selling intermediate ethanol blends at the retail level, we reviewed relevant presentations, analyses, reports, and other documents from various federal and state agencies, federal research laboratories, and industry associations, including the American Petroleum

Institute and the National Association of Convenience Stores. We also selected a nonprobability sample of knowledgeable stakeholders to interview using the same "snowball sampling" technique described for our first objective. In particular, we asked these stakeholders, among other things, to discuss the challenges, if any, associated with selling intermediate ethanol blends at the retail level. We also asked these stakeholders to identify key studies and other knowledgeable stakeholders on this topic. Over the course of our work, we interviewed officials from the following federal laboratories and agencies: DOE National Renewable Energy Laboratory (NREL), DOE Oak Ridge National Laboratory (ORNL), DOE Office of the Biomass Program, DOE Office of Vehicle Technologies Program, EPA Office of Research and Development, EPA Office of Transportation and Air Quality, EPA Office of Underground Storage Tanks, the Department of Labor's Occupational Safety and Health Administration, the National Institute of Standards and Technology, USDA Agricultural Research Service, and USDA Economic Research Service. We also interviewed state officials from the California Air Resources Board, the Minnesota State Fire Marshal Division, Northeast States for Coordinated Air Use Management,[54] and the Office of North Carolina State Fire Marshal. We interviewed representatives from the following industry associations: Growth Energy, the Renewable Fuels Association, the American Petroleum Institute, the National Association of Convenience Stores, the Society of Independent Gasoline Marketers of America, the National Association of Truck Stop Operators, the Petroleum Marketers Association of America, and the National Petrochemical and Refiners Association. Finally, we interviewed stakeholders from the following research and standards development organizations: ASTM International, Sierra Research, Inc., and Underwriters Laboratories (UL). We also conducted site visits to the research centers responsible for coordinating federal studies on the effects of intermediate ethanol blends on materials and components used in retail fuel storage and dispensing equipment. Specifically, we visited NREL facilities in Golden, Colorado; and ORNL facilities near Knoxville, Tennessee. During these site visits, we interviewed researchers conducting studies on the effects of intermediate ethanol blends on materials and components used in retail fuel-storage and -dispensing equipment. We asked these researchers to discuss available test results and the status of their testing efforts for these studies. We also toured some of the research facilities where testing was being conducted for these studies.

To examine research by federal agencies into the effects of intermediate ethanol blends on the nation's automobiles and nonroad engines, we reviewed relevant presentations, analyses, reports, and other documents from various federal and state agencies; NREL; ORNL; and industry associations, including the American Coalition for Ethanol, the National Marine Manufacturers Association, and the Outdoor Power Equipment Institute. In addition, we reviewed relevant studies and reports from academic groups and private research organizations, including the Coordinating Research Council, Inc., Minnesota State University, Mankato; and the Rochester Institute of Technology. We also selected a nonprobability sample of knowledgeable stakeholders to interview using the same "snowball sampling" technique described for our first objective. In particular, we asked these stakeholders, among other things, to identify research by federal agencies and others into the effects of intermediate ethanol blends on the nation's automobiles and nonroad engines. Over the course of our work, we interviewed officials from the following federal agencies and laboratories: DOE Office of Vehicle Technologies Program, NREL, ORNL, EPA Office of Research and Development, and EPA Office of Transportation and Air Quality. We also interviewed state officials from the California Air Resources Board and Northeast States for Coordinated Air Use Management. We interviewed representatives from the following industry associations: the American Petroleum Institute, Growth Energy, the Renewable Fuels Association, the Alliance of Automobile Manufacturers, the Association of International Automobile Manufacturers, Inc.,[55] the Outdoor Power Equipment Institute, the Engine Manufacturers Association, the National Marine Manufacturers Association, and the International Snowmobile Manufacturers Association. Finally, we interviewed stakeholders from the following academic and research organizations: the Coordinating Research Council, Inc.; the Rochester Institute of Technology; and Minnesota State University, Mankato. We also conducted site visits to the research centers responsible for coordinating federal studies on the effects of intermediate ethanol blends on automobiles and nonroad engines. Specifically, we visited NREL facilities in Golden, Colorado; and ORNL facilities near Knoxville, Tennessee. We also visited a private research facility in Aurora, Colorado, where some of the automobile testing for federal studies has taken place. During these site visits, we interviewed researchers conducting studies on the effects of intermediate ethanol blends on automobiles and nonroad engines. We asked these researchers to discuss available test results and the status of their testing efforts for these studies. We also toured some of the research facilities where testing was being conducted for these studies. Due to ongoing litigation over EPA's decision to allow ethanol blends with 15 percent ethanol (E15) for use with certain automobiles, we did not evaluate any research by federal agencies and others into the effects of intermediate ethanol blends on automobiles and nonroad engines.

We conducted this performance audit from April 2010 to June 2011, in accordance with generally accepted government auditing standards. Those standards require that we plan and perform the audit to obtain sufficient, appropriate evidence to provide a reasonable basis for our findings and conclusions based on our audit objectives. We believe that the evidence obtained provides a reasonable basis for our findings and conclusions based on our audit objectives.

APPENDIX II.

STUDIES BY NONFEDERAL ORGANIZATIONS ON THE EFFECTS OF INTERMEDIATE ETHANOL BLENDS IN AUTOMOBILES

Nonfederal organizations are conducting research on the effects of intermediate ethanol blends in automobiles. For example, in addition to the research the Coordinating Research Council, Inc. (CRC) is conducting, in coordination with DOE and EPA, it has both ongoing and completed research projects on a range of related topics, including evaporative and exhaust emissions for various intermediate ethanol blends. A CRC representative told us that it expects to complete these projects by early 2012. Based on this research, CRC has published 10 reports as of March 2011 (see table 3).

Two academic organizations have also conducted research on intermediate ethanol blends in automobiles. Specifically, the Minnesota Center for Automotive Research at Minnesota State University, Mankato, has issued five studies looking at the effects of ethanol blends containing 20 percent ethanol (E20) on fuel system components.[56] These studies received funding from the Minnesota Department of Agriculture and appear on the department's Web site.[57] In addition, the Center for Integrated Manufacturing Studies at Rochester Institute of Technology in New York has studied the effects of E20 on automobile exhaust, drivability, and maintenance, with funding from DOT. To date, the center has published one report and expects to publish at least two more later in 2011, along with a final summary report to DOT.[58]

Table 3. Published CRC Reports on Effects of Intermediate Ethanol Blends in Automobiles

Research topic	Reports
Catalyst durability	• *Mid-Level Ethanol Blends Catalyst Durability Study Screening*, June 2009.
Drivability performance	• *2006 CRC Hot-Fuel-Handling Program*, January 2007 • *2008 CRC Cold-Start and Warmup E85 and E15/E20 Driveability Program*, October 2008 • *2010 CRC Altitude Hot-Fuel-Handling Program*, January 2011
Evaporative emissions	• *Fuel Permeation from Automotive Systems*, September 2004. • *Fuel Permeation from Automotive Systems: E0, E6, E10, E20 and E85*, December 2006. • *Vehicle Evaporative Emission Mechanisms: A Pilot Study*, June 2008 • *Enhanced Evaporative Emission Vehicles*, March 2010
Exhaust emissions	• *Effects of Vapor Pressure, Oxygen Content, and Temperature on CO Exhaust Emissions*, May 2009.
Onboard diagnostic systems	• *Impact of E15/E20 Blends on OBDII Systems – Pilot Study*, March 2010.

Source: GAO analysis of CRC information.

APPENDIX III.

COMMENTS FROM THE ENVIRONMENTAL PROTECTION AGENCY

UNITED STATES ENVIRONMENTAL PROTECTION AGENCY
WASHINGTON, D.C. 20460

MAY 20 2011

OFFICE OF
SOLID WASTE AND
EMERGENCY RESPONSE

Mr. Frank Rusco
Director, Natural Resources and Environment
Government Accountability Office
Washington, DC 20548

Dear Mr. Rusco:

Thank you for the opportunity to comment on the draft report entitled "Biofuels: Challenges to the Transportation, Sale, and Use of Intermediate Ethanol Blends (GAO-11-513)." I am responding on behalf of the Office of Solid Waste and Emergency Response (OSWER) as well as the Office of Air and Radiation (OAR). Their comments have been incorporated into this consolidated Environmental Protection Agency (EPA or Agency) response. Below are our most significant comments on the report's one recommendation for EPA and on the information provided in the report itself. Other technical comments are included in the Enclosure.

Recommendation

To reduce uncertainty about the potential environmental impacts of storing intermediate ethanol blends at retail refueling locations, we recommend that the Administrator of EPA determine what additional research is necessary to better understand the compatibility of intermediate ethanol blends with UST systems, including the compatibility of specific UST components, and develop a plan to undertake such research.

EPA agrees with the importance of ensuring that the owners and operators of underground storage tank (UST) systems, when and if they choose to move to store higher blends of ethanol, are able to demonstrate that their UST systems are compatible with the stored fuel. Efforts are underway to evaluate the suitability of current UST systems to store new fuels. One project involves evaluating the functionality of current UST leak detection technologies when used with ethanol-blended fuels. We are also working with Department of Energy labs to understand the impacts of mid-level ethanol blends on materials used in tank systems. In addition, EPA is working to assess the impact of biofuel releases to the environment, and to adapt remediation tools to account for the differences in ethanol-blended fuels.

While EPA believes that a targeted approach to research will be important to accommodate the move to higher ethanol blends, we also acknowledge that there will always be uncertainty concerning the compatibility of legacy equipment with these fuel blends. Due to a multitude of factors, including age and prior use of equipment, number of UST system components, variation of products available on the market over time, and the sheer multitude of possible configurations of UST systems, the ability to determine compatibility with the approximately 600,000 UST systems currently in use is limited.

After carefully weighing the need to fully understand the compatibility issues of the higher blends of ethanol against a realistic appraisal of the ability of research to address all the permutations of UST configurations, EPA has chosen a policy approach to provide certainty to the UST market. Those UST owners, who cannot demonstrate compatibility of their systems with the higher blends of ethanol, cannot store those fuels. We are in the final stages of developing guidance for UST owners on how to determine the compatibility of their tank systems if they wish to store higher blends of ethanol. We have had extensive conversations with UST stakeholders, including the equipment industry, states, and the regulated community in developing this guidance. This guidance will provide the certainty the industry needs to safely store higher ethanol blends, while meeting the federal requirement for compatibility and ensuring protection of human health and the environment.

Rather than developing a plan to undertake additional compatibility research, EPA will continue to work with other federal agencies, industry and other stakeholders to assist tank owners to safely transition to new fuels. We anticipate that additional, targeted research may be necessary to facilitate that transition. The Agency will consider how to best partner with these groups to advance that research.

General Comments

There is a great deal of interest in alternative fuels, prompted by federal law (i.e., the Renewable Fuel Standard), by the concern of the impact of the continued use of petroleum based fuels and by the rise in the price at the pump. It is, however, not mandatory for a tank owner to move to the intermediate blends of ethanol (such as E15). If a tank owner chooses to sell E15, and therefore store that blend in their UST system, they may need to upgrade certain components in order to ensure that their UST system is compatible. Tank owners who are unable to prove their UST systems are compatible also have the option to not store E15 – that is to continue to store E10. As E15 is only legal for use in a subset of motor vehicles, we believe there will be a continued demand for E10.

We believe most tank owners will be able to demonstrate compatibility for the major components of their UST systems, including tanks and piping, and will only need to upgrade smaller components such as the submersible pump, tank probes, seals, and gaskets. These components are typically accessible under sump covers. For that reason, the references to "excavation" in the draft report are not accurate. Further, most owners, who wish to upgrade their systems, will need to make less costly targeted upgrades to readily accessible components, at a cost substantially less than the cost of an entire tank system replacement.

In closing, we believe that providing owners and operators of UST systems clarity on the implementation of existing federal regulations with emerging biofuels, including higher blends of ethanol, is critical as we continue to transition to these fuels. It is clear that close collaboration with our federal and state partners and working with our stakeholders in industry

If you have any questions or concerns regarding our comments or response to the recommendation, EPA would be happy to meet with you prior to GAO finalizing this report. Please feel free to contact me or Mark Barolo at 703-603-7141 if there is any additional follow up required.

Sincerely,

Lisa Feldt for

Mathy Stanislaus
Assistant Administrator

Enclosure

cc: Gina McCarthy, OAR
Karl Simon, OAR
Bob Trent, OCFO
Carolyn Hoskinson, OSWER
Mark Barolo, OSWER
Linda Gerber, OSWER
Johnsie Webster, OSWER

APPENDIX IV.

COMMENTS FROM THE OCCUPATIONAL SAFETY AND HEALTH ADMINISTRATION

U.S. Department of Labor

Assistant Secretary for
Occupational Safety and Health
Washington, D.C. 20210

MAY 13 2011

Mr. Frank Rusco, Director
Natural Resources and Environment
U.S. Government Accountability Office
441 G Street, N.W.
Washington, D.C. 20548

Dear Mr. Rusco:

Thank you for the opportunity to comment on the Government Accountability Office's (GAO) proposed report, *BIOFUELS: Challenges to the Transportation, Sale, and Use of Intermediate Ethanol Blends*. OSHA appreciates the time and effort that GAO took in its evaluation of the ethanol industry.

OSHA is addressing the worker safety-related uncertainties that are attached to the complex issues surrounding biofuels and will address GAO's recommendation in depth in its Statement of Executive Action. We anticipate that the many challenges associated with the use, sale and transportation of biofuels can be addressed by our agency working in conjunction with the Environmental Protection Agency, the Department of Transportation, the Department of Energy and all other relevant organizations. We appreciate the opportunity to review and respond to GAO's draft report.

Sincerely,

David Michaels, PhD, MPH

APPENDIX V. STAFF ACKNOWLEDGMENTS

Staff Acknowledgments

Tim Minelli (Assistant Director), Nirmal Chaudhury, Cindy Gilbert, Chad M. Gorman, Jason Holliday, Michael Kendix, Ben Shouse, Barbara Timmerman, and Jack Wang made key contributions to this report.

End Notes

[1] The United States consumed about 18.8 million barrels of petroleum and petroleum products per day in 2009. The nation imported about 11.7 million barrels of petroleum and petroleum products per day in 2009—primarily crude oil but also petroleum products such as refined gasoline and jet fuel. The United States exported roughly 2 million barrels of petroleum and petroleum products per day in 2009—primarily refined products such as diesel fuel, residual fuel oil, and petroleum coke. Only 2 percent of exports were crude oil. Net imports (total imports minus exports) equaled 9.7 million barrels of petroleum and petroleum products per day in 2009.

[2] Pub. L. No. 109-58, § 1501 (2005). The act authorizes the Administrator of the EPA, in consultation with the Secretaries of Agriculture and Energy, to waive the RFS levels established in the act, by petition or on the Administrator's own motion, if meeting the required level would severely harm the economy or environment of a state, a region, or the United States, or there is an inadequate domestic supply. Throughout this report, the RFS levels established in the act are referred to as requirements, even though these levels could be waived by the Administrator.

[3] Pub. L. No. 110-140, § 201 (2007).

[4] GAO, *Biofuels: Potential Effects and Challenges of Required Increases in Production and Use*, GAO-09-446 (Washington, D.C.: Aug. 25, 2009).

[5] In this report, we use the terms "automobiles" and "motor vehicles" to refer to (1) light-duty vehicles, including passenger cars; (2) light-duty trucks, including pickup trucks, minivans, passenger vans, and sport-utility vehicles; and (3) medium-duty passenger vehicles, including large sport-utility vehicles and passenger vans.

[6] According to DOE's Office of Energy Efficiency and Renewable Energy, intermediate ethanol blends include E15 and E20 and are defined as having an ethanol content greater than 10 percent and less than 85 percent.

[7] In this report, we use the term "nonroad engines" to refer to nonroad products with gasoline engines, including (1) lawn and garden equipment, such as lawn mowers, weed trimmers, leaf blowers, chainsaws, and snowblowers; (2) recreational engines and vehicles, such as all-terrain vehicles, dirt bikes, and snowmobiles; (3) recreational marine vehicles; (4) construction and industrial equipment and vehicles, such as forklifts and paving equipment; (5) commercial equipment, such as generators and air compressors; (6) farm equipment, such as tractors and combines; and (7) logging equipment.

[8] The advanced biofuel category includes ethanol imported from some member nations of the Caribbean Basin Initiative and Brazil, which primarily use sugarcane to make ethanol.

[9] 75 Fed. Reg. 76790 (Dec. 9, 2010).

[10] Terminals on the East Coast are large integrated facilities with marine, pipeline, and tanker truck receiving and dispatching capabilities. Although some terminals have rail access, they were not originally designed to support rail as a major mode for transporting fuel.

[11] Ethanol transported for fuel is referred to as fuel-grade ethanol and typically contains 2 percent denaturant, such as gasoline, to render it unfit for human consumption.

[12] According to DOE documentation, there were more than 8 million light-duty flexible-fuel vehicles on U.S. roads as of May 2010 and 2,051 retail fueling locations offering E85 as of June 2010. Because a gallon of ethanol contains only about two-thirds the energy of a gallon of gasoline, the use of E85 results in an approximately 25 percent reduction in fuel economy.

[13] As reported in NPN, *MarketFacts 2010*, (Park Ridge, Ill., 2010) www.npnweb.com.

[14] The Nielsen Company (Washington, D.C., May 2010) www.nielsen.com.

[15] As reported in NPN's *MarketFacts 2010*.

[16] 29 C.F.R § 1910.106(g)(3)(iv).

[17] UL is a standards development organization that certifies (e.g., tests and approves) equipment based on standards it develops. According to OSHA officials, two other laboratories—CSA International and Intertek Testing Services NA, Inc.—also certify dispensing equipment based on UL's standards. However, representatives from these two laboratories told us that they are currently conducting little, if any, certification activities for dispensing equipment.

[18] According to EPA and OSHA officials, OSHA's requirements for dispensing equipment and EPA's requirements for UST systems overlap at the submersible turbine pump, which delivers fuel from the UST to the dispenser. Therefore, along with meeting EPA's compatibility requirements, these pumps must also be certified for safety by a nationally recognized testing laboratory, such as UL, per OSHA requirements. OSHA also has compatibility requirements for UST systems, but unlike its requirements for dispensing equipment, OSHA does not require UST equipment to be certified by a nationally recognized testing laboratory.

[19] The mission of the international nonprofit National Fire Protection Association is to reduce the worldwide burden of fire and other hazards on the quality of life by providing and advocating consensus codes and standards, research, training, and education. The International Code Council is a membership association dedicated to building safety and fire prevention. The council develops the codes and standards used to construct residential and commercial buildings, including homes and schools.

[20] In a carbureted fuel system, the air-to-fuel ratio is preset at the factory based on the expected operating conditions of the engine such as ambient temperature, atmospheric pressure, speed, and load.

[21] EPA, *Renewable Fuel Standard Program (RFS2) Regulatory Impact Analysis*, EPA-420-R-10-006 (Washington, D.C., February 2010).

[22] EPA used three control cases—high-ethanol, primary or mid-ethanol, and low-ethanol—to account for different levels of projected cellulosic biofuel production. EPA then compared each of its control cases against a "reference" case based on estimates made by the Energy Information Administration in its 2007 *Annual Energy Outlook* for ethanol production by 2022. EPA focused on scenarios in which ethanol consumption increased greatly in all 50 states. While not discussed in EPA's report, an additional option would be increased use of E85, primarily in the Midwest. However, additional E85 fueling stations in the Midwest would be needed for this option.

[23] According to EPA's analysis, there is significant uncertainty regarding its estimate for the production of cellulosic biofuels by 2022.

[24] GAO, *Freight Transportation: National Policy and Strategies Can Help Improve Freight Mobility*, GAO-08-287 (Washington, D.C.: Jan. 7, 2008).

[25] According to company representatives, Kinder Morgan invested approximately $10 million to modify its Central Florida Pipeline for ethanol shipments, which included chemically cleaning the pipeline, replacing equipment that was incompatible with ethanol, and expanding storage capacity at its Orlando terminal.

[26] This research can be found at http://primis.phmsa.dot.gov/matrix/ after typing "ethanol" into the search feature.

[27] This research can be found at http://primis.phmsa.dot.gov/matrix/ after typing "ethanol" into the search feature.

[28] Since the February 2008 announcement, Buckeye has discontinued its role in the proposal. In March 2009, Magellan and POET signed a joint development agreement to continue assessing the feasibility of a dedicated 1,700 mile pipeline moving ethanol from the Midwest to the major Northeastern markets. Pipeline costs were estimated to exceed $3.5 billion. A revised press release, issued in January 2010, increased the estimated length of the pipeline to 1,800 miles and the cost estimate to $4 billion.

[29] DOE, *Report to Congress: Dedicated Ethanol Pipeline Feasibility Study* (Washington, D.C., March 2010).

[30] EPA, *Testimony of Lisa Jackson, Administrator, U.S. Environmental Protection Agency, before the Committee on Agriculture, United States House of Representatives* (Washington, D.C., Mar. 10, 2011).

[31] 75 Fed. Reg. 68044 (Nov. 4, 2010).

[32] Sierra Research, Inc., *Identification and Review of State/Federal Legislative and Regulatory Changes Required for the Introduction of New Transportation Fuels*, Report No. SR2010-08-01 (Sacramento, Calif., Aug. 4, 2010), prepared for the American Petroleum Institute.

[33] These standards cover blends with up to 25 percent ethanol (E25). UL published safety standards for certifying dispensing equipment for blends up to E85 in October 2007.

[34] UL certified dispensers from two manufacturers in March 2010 for use with blends up to E25, and in June 2010 for blends up to E85. According to knowledgeable OSHA officials, if employees covered by OSHA used or worked on unapproved equipment dispensing higher ethanol blends, it would likely constitute a violation of OSHA requirements. However, these officials said that OSHA is not aware of any complaints, referrals, or notifications of serious accidents involving this equipment.

[35] According to DOE officials, this research used a testing fluid containing 17 percent ethanol, acids, water, and minerals to represent worst-case scenarios for fuel. See UL, *Dispensing Equipment Testing With Mid-Level Ethanol/Gasoline Test Fluid* (Washington, D.C., November 2010), prepared for DOE.

[36] This research was on the exposure of common dispenser materials to testing fluids containing 17 and 25 percent ethanol, plus acids, water, and minerals. See ORNL, *Intermediate Ethanol Blends Infrastructure Materials Compatibility Study: Elastomers, Metals, and Sealants*, ORNL/TM-2010/326 (Oak Ridge, Tenn., March 2011). According to DOE, elastomers are a class of polymers widely used in fuel dispenser systems as o-rings and gasket-type seals.

[37] According to knowledgeable EPA officials, it is possible to purchase a new UST system meeting EPA requirements for compatibility with all ethanol blends. However, according to EPA officials, most tank owners still purchase some components that are only approved for use with E10.

[38] 75 Fed Reg. 70241 (Nov. 17, 2010).

[39] ORNL, ORNL/TM-2010/326.

[40] DOE recently estimated that modifying fuel pumps to make them compatible with E15 should cost $1,000 or less per pump, depending on pump-specific variables. See DOE, *Statement of Dr. Henry Kelly, Acting Assistant Secretary For Energy Efficiency, U.S. Department of Energy, Before the Committee on Environment and Public Works, United States Senate* (Washington, D.C., Apr. 13, 2011).

[41] EPA, Office of Transportation and Air Quality, EPA-420-R-10-006.

[42] EPA officials told us that it does not collect data on tank configurations at retail locations. As a result, we relied on information from industry representatives to illustrate this potential challenge.

[43] Several UL representatives told us that the announcement did not mean that UL was recertifying existing equipment for use with intermediate blends or that existing equipment could be used with E15 (because the ethanol content of a specific blend, like E15, may vary and potentially could exceed 15 percent under normal business conditions).

[44] 75 Fed. Reg. 68044 (Nov. 4, 2010).

[45] 75 Fed. Reg. 12470 (Mar. 16, 2010).

[46] See ORNL, *Technical Issues Associated with the Use of Intermediate Ethanol Blends (>E10) in the U.S. Legacy Fleet: Assessment of Prior Studies* (Oak Ridge, Tenn., August 2007).

[47] CRC is a nonprofit organization supported by the petroleum and automotive equipment industries. CRC operates through committees made up of technical experts from industry and government who voluntarily participate.

[48] Of this total amount, DOE has provided about $45 million, including less than $65,000 for NREL and ORNL to review studies conducted by the Minnesota Center for Automotive

Research at Minnesota State University, Mankato; and the Center for Integrated Manufacturing Studies at Rochester Institute of Technology. These efforts are not included in table 2.

[49] EPA regulates the emissions of air pollutants—which are known or reasonably anticipated to endanger public health or welfare—from mobile sources such as automobiles. These pollutants include hydrocarbons (such as benzene and acetaldehyde), carbon monoxide, nitrogen oxides, and volatile organic compounds.

[50] NREL, ORNL, *Effects of Intermediate Ethanol Blends*.

[51] GAO-09-446.

[52] The information gathered from these interviews cannot be used to generalize findings or make inferences about the entire population of knowledgeable stakeholders on intermediate ethanol blends and related topics. Although the sample provides some variety, it is unlikely to capture the full variability of knowledgeable stakeholders, and it cannot provide comprehensive insight into the views of any one group of knowledgeable stakeholders. This is because, in a nonprobability sample, some elements of the population being interviewed have no chance, or an unknown chance, of being selected as part of the sample. However, the information gathered during these interviews allows us to discuss various stakeholder views on intermediate ethanol blends, and it provides important context overall. It also helps us interpret the documentation and other testimonial evidence we have collected.

[53] National Commission on Energy Policy, *Task Force on Biofuels Infrastructure*; EPA-420-R-10-006; and DOE, *Report to Congress: Dedicated Ethanol Pipeline Feasibility Study*.

[54] Northeast States for Coordinated Air Use Management is an association of the state air quality agencies from Connecticut, Maine, Massachusetts, New Hampshire, New Jersey, New York, Rhode Island, and Vermont.

[55] The Association of International Automobile Manufacturers, Inc. is now known as the Association of Global Automakers, Inc.

[56] Bruce Jones, et al., *The Effects of E20 on Elastomers Used in Automotive Fuel System Components*, (Mankato, Minn., Feb. 22, 2008); Bruce Jones, et al., *The Effects of E20 on Plastic Automotive Fuel System Components*, (Mankato, Minn., Feb. 21, 2008); Bruce Jones, et al., *The Effects of E20 on Metals Used in Automotive Fuel System Components*, (Mankato, Minn., Feb. 22, 2008); Nathan Hanson, et al., *The Effects of E20 on Automotive Fuel Pumps and Sending Units,* (Mankato, Minn., Feb. 21, 2008); and Gary Mead, et al., *An Examination of Fuel Pumps and Sending Units During a 4000 Hour Endurance Test in E20*, (Mankato, Minn., Mar. 25, 2009).

[57] *E20 Test Results*, in the Minnesota Department of Agriculture database, http://www.mda.state.mn.us/en/renewable/ethanol/e20testresults.aspx (accessed Apr. 4, 2011).

[58] B. Hilton and B. Duddy, "The Effect of E20 Ethanol Fuel on Vehicle Emissions," *Proceedings of the Institute of Mechanical Engineers, Part D: Journal of Automobile Engineering*, vol. 223 no. 12 (2009).

In: Biofuel Use in the U.S.
Editors: S. Alonso and M. R. Ortega
ISBN: 978-1-62100-441-7

Chapter 3

THE MARKET FOR BIOMASS-BASED DIESEL FUEL IN THE RENEWABLE FUEL STANDARD (RFS)

Brent D. Yacobucci

SUMMARY

The market for biomass-based diesel (BBD) fuel, most notably biodiesel, has expanded rapidly since 2004, largely driven by federal policies, especially tax credits and a mandate for their use under the federal Renewable Fuel Standard (RFS). Most expect that the majority of the BBD fuel quota in the RFS will be met using biodiesel produced from soybean oil. Biodiesel from other feedstocks, and other biomass-based substitutes (e.g., synthetic diesel from cellulosic feedstocks or algae) could play a larger role in the future, although currently these other alternatives are prohibitively expensive to produce in sufficient quantities.

Biodiesel production remains expensive relative to conventional petroleum-based diesel (even with tax credits), largely due to the reliance on soybean oil (a relatively expensive commodity) as a feedstock. Biodiesel and other BBD fuel production remains dependent on both tax incentives and the RFS mandates, as evidenced by a drop in production from 2009 to 2010. The expiration of the BBD tax credits after 2009 more than counteracted the increase in the RFS mandate from 2009 to 2010. Whether enough biodiesel production capacity will come online in 2011 to meet an even larger mandate remains to be seen.

The absence of the tax incentive for most of 2010, along with high soybean oil prices, caused 2010 biodiesel production to drop significantly—to the point that 2010 production may be below that needed to meet the RFS mandate. Any shortfall in supply for the 2010 BBD mandate may be met using credits generated in 2011, leading to even tighter markets going forward. These credits—referred to as RINs (Renewable Identification Numbers)—may be used by fuel suppliers to meet their obligations, banked for the next calendar year, or traded to other entities. In this way, analysis of the financial market for RINs may serve as a useful method for evaluating the overall market for BBD fuels. As BBD RINs become scarce, their price has increased dramatically (by nearly an order of magnitude over the past year). At some point, the value of the RINs may increase enough to bring idled production capacity back online or promote an increase in new capacity development and/or imports, especially if BBD producers expect RIN prices to stay high in the future.

This report discusses the current market for BBD fuels and their corresponding RINs under the RFS. It examines the role that the RIN market may play as an economic incentive for the production of biodiesel and other BBD fuels in the future. Lessons learned from the experience with the BBD quota and the associated RIN market may provide insights into the future RIN markets for other advanced biofuels and perhaps for the RFS as a whole.

INTRODUCTION

Increasing dependence on foreign sources of crude oil, concerns over global climate change, and the desire to promote domestic rural economies have raised interest in renewable biofuels as an alternative to petroleum in the U.S. transportation sector. In response to this interest, U.S. policymakers have enacted an increasing variety of policies to directly support U.S. biofuels production and use.[1] Policy measures include blending and production tax credits to lower the cost of biofuels to end users, research grants to stimulate the development of new biofuels technologies, and loans and loan guarantees to facilitate the development of biofuels production and distribution infrastructure. Perhaps most important, Congress established minimum usage requirements to guarantee a market for biofuels irrespective of their cost.[2]

This guaranteed market—the Renewable Fuel Standard (RFS)—requires refiners and other fuel suppliers to include an increasing amount of biofuels in transportation fuel.[3] By 2022, the RFS requires the use of 36 billion gallons of renewable fuel. Within the larger mandate, there are sub-mandates ("carve-outs") for the use of advanced biofuels (fuels other than corn starch ethanol with at least 50% lower lifecycle greenhouse gas emissions than conventional fuels).[4] Within the advanced biofuel quota, there are additional sub-mandates for cellulosic biofuels and for biomass-based diesel substitutes (BBD).

In the near term, most expect that the majority of the BBD quota (and a significant portion of the larger advanced biofuel quota) will be met using biodiesel produced from soybean oil, although biodiesel from other feedstocks, as well as other biomass-based substitutes (e.g., synthetic diesel from cellulosic feedstocks or algae), could play a larger role in the future.

One of the key incentives, a $1.00-per-gallon tax incentive for the production and/or use of biodiesel and renewable diesel (a related type of BBD), expired at the end of 2009. This tax incentive was eventually extended retroactively for 2010 through 2011 in December 2010. However, the absence of this incentive for most of 2010, along with high soybean oil prices, caused 2010 biodiesel production to drop significantly—to the point that 2010 production is expected to be below that needed to meet the RFS mandate.

Fuel suppliers comply with the RFS by submitting to the Environmental Protection Agency (EPA) sufficient credits (Renewable Identification Numbers, or RINs) to cover their annual obligations. RINs may be used by fuel suppliers to meet their obligations, banked for the next calendar year, or traded to other entities. As BBD RINs have become scarce, their price has increased dramatically (by an order of magnitude over the past year). Analysis of the financial market for RINs may serve as a useful method for evaluating the overall market for BBD fuels. At some point, the value of the RINs may increase enough to bring idled production capacity back online or to spur new capacity development, especially if BBD producers expect RIN prices to stay high in the future. Some of the potential 2010 BBD shortfall may be met using RINs generated in 2011, although this added demand for 2011 RINs could push their price even higher.

This report discusses the current market for BBD fuels and their corresponding RINs under the RFS. It examines the role that the RIN market may play as an economic incentive for the production of biodiesel and other BBD fuels in the future. Lessons learned from the BBD quota and the associated RIN market may provide insights into the future RIN markets for other advanced biofuels and perhaps for the RFS as a whole.

WHAT IS BIOMASS-BASED DIESEL (BBD)?

Biomass-based diesel (BBD) fuel is a term used for any type of biologically derived diesel fuel for compliance with the RFS. BBD is an umbrella term that captures a range of fuels, including biodiesel, renewable diesel, synthetic (Fischer-Tropsch) diesel, algae-based diesel fuel, and cellulosic diesel.

The RFS mandates the use of BBD substitutes. While this is often referred to as a "biodiesel mandate," the term biodiesel refers to a fuel produced through a specific process. The RFS does not distinguish between different BBD fuels or the process by which they are produced, as long as they meet requirements for greenhouse gas emissions reductions.[5] For clarity, when referring to the larger class of fuels, this report uses the term biomass-based diesel, or BBD; the term "biodiesel" is used when referring specifically to fuel that meets the statutory definition of biodiesel.

DIFFERENT TYPES OF BIOMASS-BASED DIESEL

Biodiesel: A diesel fuel substitute (mono-alkyl esters) produced through a chemical conversion process (transesterification). Production feedstocks generally include virgin vegetable/animal oils or recycled restaurant grease and an alcohol (generally methanol). Meets ASTM Standard D6751.

Renewable Diesel: A diesel fuel substitute produced through a thermochemical process and chemically distinct from biodiesel (i.e., non-ester fuel). Feedstocks include animal or vegetable oils but (unlike biodiesel) do not include alcohol inputs. Meets ASTM Standard D975 or D396.

Fischer-Tropsch (FT) Diesel: Often referred to as biomass-to-liquids (BTL), biomass-based FT diesel is produced through a two-step process where the feedstock is converted to a synthesis gas (syngas) and then synthetic diesel fuel is produced through a catalytic process. FT synthesis can also be used to produce natural gas-to-liquids (GTL) or coal-to-liquids (CTL) fuels.

Algae-Based Biofuels: Algae has the potential to produce much higher per-acre yields of oils (lipids) than conventional crops. Algae can potentially be used to produce any of the above diesel fuel substitutes depending on the production process.

Cellulosic Diesel: Instead of using vegetable or other oils as a feedstock, cellulosic material (woody or fibrous plant material) could be used to develop a variety of biofuels, including BBD fuels. Cellulose could be converted to BBD through FT synthesis or through other thermal or chemical pathways.

THE U.S. BIOMASS-BASED DIESEL INDUSTRY

Domestic Production

In the United States, BBD production is dominated by biodiesel, mostly from virgin soybean oil, although other feedstocks (e.g., animal fats, canola, recycled grease) play a role, as does non-biodiesel fuel (e.g., renewable diesel). Data on U.S. biodiesel production are limited, especially compared to other energy and agricultural commodities, and data on non-biodiesel BBD are virtually nonexistent. However, despite relatively little detailed data on biodiesel and other BBD markets, one fact is clear: biodiesel production and consumption in the United States have grown dramatically in the last several years, driven mostly by a tax credit for the production/blending of biodiesel and renewable diesel and by mandates in the RFS. (See Figure 1.) In 2004, before the biodiesel tax credit was established (2005), production and consumption remained below 30 million gallons annually. By 2008, both production and consumption had increased more than tenfold. Even after the peak in 2008 and subsequent drop, production and consumption levels remain well above pre-2005 levels.

Biodiesel production plants are generally located in the Midwest where soybean and other oilseed production is concentrated, or in Texas close to large supplies of animal fats. (See Figure 2.) Six states—Texas, Iowa, Illinois, Indiana, Minnesota, and Missouri—accounted for 60% of production capacity in 2009. In December 2009 the U.S. Department of Energy's Energy Information Administration (EIA) estimated that there was roughly 2 billion gallons in annual biodiesel production capacity.[6] However, plants capable of producing biodiesel may produce a variety of other products as well (or instead), and thus actual biodiesel production tends to be far below theoretical capacity: while most chemical plants operate at 80%-90% nameplate capacity, 2009 biodiesel capacity utilization was around 33%. Even at the peak in 2008, production was well below potential capacity.

As noted above, two key domestic incentives—tax credits and the RFS mandate—have driven expansion of domestic production and consumption of BBD. The American Jobs Creation Act of 2004 established a tax credit of $1.00 per gallon for the production and/or blending of biodiesel.[7] Since that time, BBD tax credits have been expanded to include credits for the production of renewable diesel and a Small Agri-Biodiesel[8] Producer Credit.[9] (See Table 1.) The RFS mandate for BBD fuels increases to 1.0 billion gallons by 2012, which would lead to production levels nearly 50% higher than the 2008 peak and domestic consumption levels nearly three times higher than the 2007 peak. Total production capacity, capacity utilization at biodiesel plants, and/or imports would need to expand dramatically in the next few years to meet that mandate.

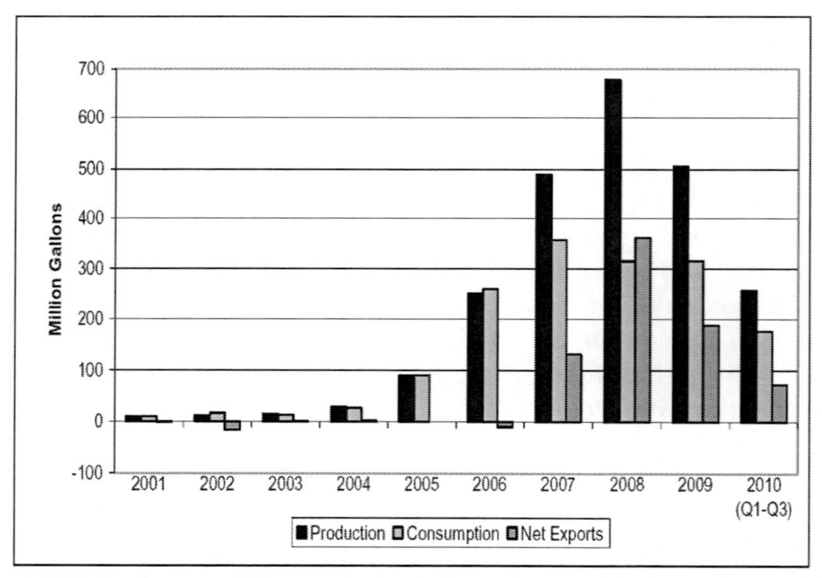

Source: U.S. Department of Energy, Energy Information Administration, October 2010 Monthly Energy Review, Washington, DC, October 28, 2010, Table 10.4, http://www.eia.doe.gov/mer/.

Notes: Domestic production dropped after 2008 partly due to a change in the biodiesel tax credit, as well as the establishment of a European Union biodiesel import tariff. Production further dropped from 2009 to 2010 as the biodiesel tax credit expired after 2009 and was not retroactively extended until late in 2010.

Figure 1. Annual U.S. Biodiesel Production, Consumption, and Exports.

Million Gallons per Year (mgy), by State

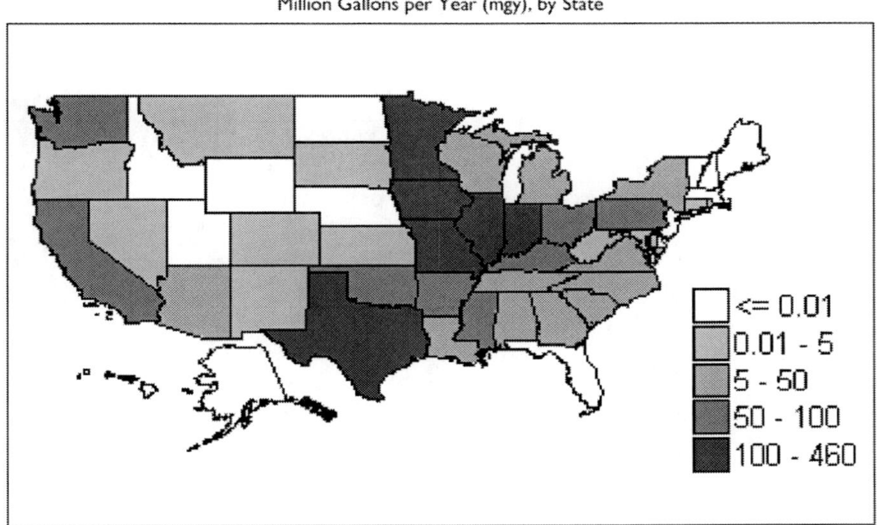

Source: CRS analysis of data from Energy Information Administration, Biodiesel Monthly Report, Washington, DC, October 2010, Table 4, *http://www.* eia. doe. gov/cneaf/solar.renewables/page/biodiesel/biodiesel.html.

Notes: Biodiesel plant capacity utilization (about 33%) is far below capacity utilization for other chemical plants (generally 80%-90%). Thus, total capacity far exceeds actual biodiesel production.

Figure 2. 2009 Biodiesel Production Capacity.

Biodiesel Markets

Over the past roughly six years (since the enactment of the tax credit), biodiesel prices have generally followed the same trend as conventional diesel prices, but have remained consistently higher despite the tax credit. (See Figure 3.) Between September 2005 and July 2010, the spread between biodiesel and conventional diesel prices ranged from $0.26 to $1.11 per gallon. Exacerbating the price spread is that biodiesel has a slightly lower energy content per gallon than diesel fuel. Thus, the per-mile cost difference between biodiesel and conventional diesel is slightly higher than the per-gallon prices would indicate. However, users may see benefits from using biodiesel despite its higher costs, including the potential for lower pollutant emissions and the use of domestically produced fuel.

Table 1. Federal Biomass-Based Diesel Tax Incentives

Tax Incentive	Established	Description	Scheduled Termination
Biodiesel Tax Credit	2004 by P.L. 108-357	Producers of biodiesel or diesel/biodiesel blends may claim a tax credit of $1.00 per gallon of biodiesel	End of 2011
Small Agri-Biodiesel Producer Credit	2005 by P.L. 109-58	An agri-biodiesel (biodiesel produced from virgin agricultural products) producer with less than 60 million gallons per year in production capacity may claim a credit of 10 cents per gallon on the first 15 million gallons produced in a year	End of 2011
Renewable Diesel Tax Credit	2005 by P.L. 109-58	Producers of renewable diesel or diesel/renewable diesel blends may claim a credit of $1.00 per gallon of renewable diesel	End of 2011
Cellulosic Biofuel Production Credit	2008 by P.L. 110-246	Producers of cellulosic biofuel (including cellulosic diesel) may claim a credit of $1.01 per gallon	End of 2012

Source: CRS Report R40110, *Biofuels Incentives: A Summary of Federal Programs*, by Brent D. Yacobucci.

One primary reason for the high price of biodiesel is that the key feedstock—soybean oil—is a relatively expensive commodity that tends to keep pace with petroleum (and conventional diesel) prices. It takes roughly one gallon of soybean oil to produce a gallon of biodiesel (along with other inputs),[10] and per-gallon soybean oil prices have been on par with per-gallon diesel prices in recent years. (See Figure 3.) The relationship between conventional diesel prices and soybean oil and biodiesel prices is particularly strong: the correlation between conventional diesel prices on one hand and either soybean oil or biodiesel on the other is 84% and 89%, respectively.[11] Adding production process, transportation, and distribution costs thus raises the price of soy biodiesel above that of conventional diesel, even when the value of the biodiesel tax credit is included.

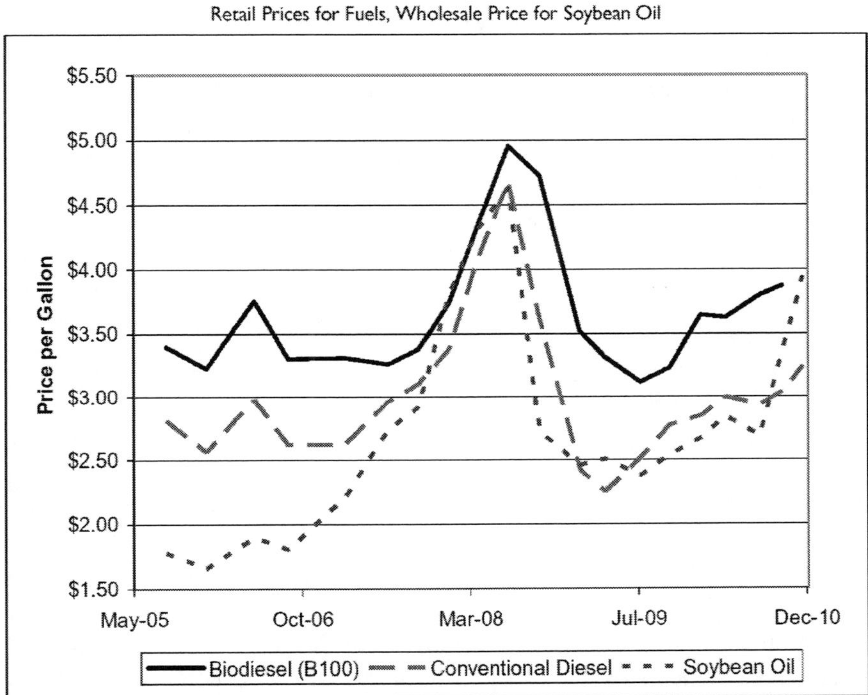

Source: Biodiesel and Conventional Diesel prices: Department of Energy, *Clean Cities Alternative Fuel Price Report*, Landover, MD, September 2005 to October 2010, http://www.afdc.energy.gov/afdc/price_report.html. Soybean oil prices: USDA, Foreign Agricultural Service, *Oilseeds: World Markets and Trade Monthly Circular*, January 2011, http://www.fas.usda.gov/psdonline/circulars/oilseeds.pdf.

Note: Fuel prices are retail prices, including taxes/tax credits. Soybean oil prices are wholesale (Decatur) prices. Soybean oil prices are converted from $/pound to $/gallon based on 7.5 lbs per gallon. Biodiesel data are through October 2010; soybean oil and diesel data are through December 2010. The dramatic run-up in soybean oil prices (and the less dramatic but still noticeable increase in conventional diesel prices) in late 2010 could lead to similar increases in biodiesel prices.

Figure 3. Comparison of Biodiesel, Conventional Diesel, and Soybean Oil Prices.

The prices for most agricultural commodities, including soybean oil, rose dramatically in late 2010; petroleum and conventional diesel prices have also risen, although not as dramatically. It seems likely that retail biodiesel prices for November and December 2010 (which are not yet available) will reflect these higher input costs.

The development of BBD fuels that are less dependent on traditional commodity feedstocks (e.g., soy) may be necessary to decouple BBD prices from the conventional fuels they are designed to replace. Thus research is continuing in developing processes to make BBD from cellulose and algae. In many cases, cellulose is a waste product with little monetary value, while there is potential to grow algae in non-agricultural areas limiting competition for prime farmland. However, these fuels are currently significantly more expensive to produce than biodiesel from soybean oil.

RENEWABLE IDENTIFICATION NUMBERS (RINS)

Under the RFS, the mandated volumes that fuel suppliers must meet are administered through a system of credits called Renewable Identification Numbers (RINs), which correspond to actual gallons of biofuel blended or sold. A RIN is a 38-digit number assigned to a batch of renewable fuel when it is produced.[12] RINs can be seen as analogous to retail barcodes. When a fuel supplier (generally a refiner or terminal operator) blends biofuels into conventional fuel or otherwise introduces the biofuels into commerce, the RIN is "detached." Once RINs are detached, the owner of the RINs may use them to meet the current-year obligation, save them for use in the coming year (with certain limits), or trade them to another party.

Use of RINs

After each calendar year, fuel suppliers must submit to EPA sufficient RINs to cover that year's mandate. For example, if a refiner had an obligation under the mandate of 1 million gallons for 2010, by February 28, 2011, that refiner would need to submit 1 million RINs to EPA. In general, up to 20% of a given year's obligation may be met using RINs from the previous year ("RIN rollover"), while no RINs may be used from two or more years earlier.[13] Thus, RINs effectively have a one-year "shelf life." If an obligated party falls short of its obligation in a given year, it may carry that deficit forward for one year, but then may not carry any deficit forward in the following year. In that way, fuel suppliers have a limited ability to borrow RINs from the future.

Under the RFS, fuel suppliers actually face four distinct but linked RIN requirements. These requirements correspond to the four fuel types required by the RFS: (1) cellulosic biofuel; (2) biomass-based diesel (BBD) fuel; (3) advanced biofuel; and (4) overall renewable fuel. Some fuels may be used to meet different obligations simultaneously. For example, a cellulosic biofuel or BBD RIN may be used to meet the cellulosic/BBD requirement (no. 1 and/or no. 2),[14] the advanced biofuel requirement (no. 3), and the overall RFS (no. 4), while a corn-based ethanol RIN can only be applied to the overall mandate (no. 4).[15] Thus, RIN markets will likely develop separately for various classes of fuels.[16] As the cellulosic biofuel and BBD requirements are more specialized, and the fuels are generally more expensive to produce, those RINs can be expected to be more expensive most of the time, especially if there is any expected shortage in the market.

Biodiesel RIN Market

The RFS was significantly expanded by the Energy Independence and Security Act of 2007 (EISA, P.L. 110-140), including the establishment of the BBD mandate. However, while EISA mandated the use of 0.5 billion gallons of BBD in 2009, EPA did not finalize rules for the expanded RFS (often referred to as "RFS2") until February 2010.[17] Therefore, EPA combined the 0.5-billion-gallon 2009 mandate with the 0.65-billion-gallon mandate in 2010 for a combined 2009/2010 mandate of 1.15 billion gallons. EPA estimates that, counting 2008 and 2009 RINs that have been registered with the agency, an additional 345 million gallons of biodiesel in 2010 will be needed to meet the combined requirement.[18]

The Energy Information Administration (EIA) estimates that roughly 320 million gallons of biodiesel was consumed in 2009. (See **Figure 1**.) If EIA consumption estimates for the first three quarters of 2010 are accurate, then total 2010 consumption will likely fall below that in 2009.[19] Thus, supply for 2010 biodiesel RINs will almost certainly be tight, and demand may actually exceed supply.

For 2011, EISA mandates the use of 800 million gallons of biomass-based diesel. Therefore, as 2010 demand may very well exceed supply, 2011 supply will need to increase dramatically to meet the anticipated demand, and the value of any 2010 rollover RINs could increase as well.[20]

This tight RIN supply appears evident in RIN markets. Since the start of 2010, biodiesel RINs produced in 2010 have steadily increased in price, from roughly $0.10 per gallon in January 2010 to nearly $0.90 in January 2011;

RINs produced in 2011 were trading at comparable prices. (See Figure 4.) Current 2010 and 2011 RIN prices are an order of magnitude higher than 2008 and 2009 biodiesel RINs (or ethanol RINs). Unless domestic production increases dramatically or there is a large increase in imports, RIN prices will likely continue to increase through 2011 unless EPA grants a waiver from the 2011 BBD mandate, which the agency argues may be unnecessary:

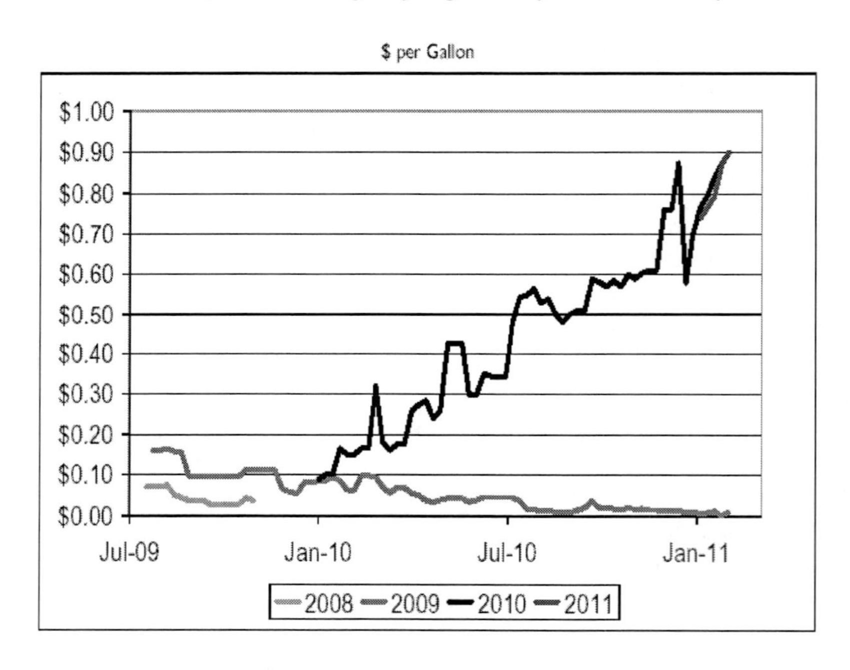

Source: "RIN Quotes," *The Ethanol Monitor*, July 20, 2009 to February 7, 2011.

Notes: RINs may be used in the year they were generated (vintage Year) or one year afterward. However, only 20% of an obligated party's requirement may be met using the prior year's RINs. After the settlement period (February 28), RINs from two years prior are unusable and should have no market value.

Figure 4. Biodiesel RIN Prices, by Vintage Year.

To avoid a major shortfall in 2011 RIN supply, idled or new capacity will likely need to be brought online relatively quickly, capacity utilization at operating plants will need to increase dramatically, and/or imports will need to increase. The 800-million-gallon 2011 mandate exceeds U.S. production at its 2008 peak (roughly 700 million gallons). The potential effects of such a rapid increase in biodiesel production on soybean oil demand and soybean prices are unclear—doubling biodiesel production in the short run would likely lead to a doubling of soybean oil use for biodiesel production. In 2009, roughly 2 billion pounds of soybean oil were used for biodiesel production,[22] representing roughly 10% of U.S. soybean oil supply.[23] Increasing soybean oil use for biodiesel from 10% to 20% of supply in one year could have dramatic effects on soybean oil prices for biodiesel and for other uses.

> In addition to current production rates, the biodiesel industry's production potential also supports a finding that it can more than satisfy the applicable volume of biomass based diesel specified in the statute for 2011. In July of 2010, over 1.8 billion gallons of production capacity had been registered under the RFS2 program. As of September 2010, the aggregate production capacity of biodiesel plants in the U.S. was estimated at 2.6 billion gallons per year across approximately 170 facilities. Indications from the biodiesel industry are that idled facilities can be brought back into production with a relatively short leadtime. Imports of biodiesel from foreign countries also has the potential to increase the volume available for consumption in the U.S.[21]

TAX CREDIT EFFECTS

The tax credit for the production and use of BBD fuels has been a major factor in their growth over the past few years. While the RFS mandates could potentially provide a larger incentive—in the form of a guaranteed and growing market—biodiesel production has remained largely dependent on the $1.00-per-gallon tax credit. Despite a growing RFS mandate, the expiration of the tax incentives at the end of 2009 led to a drop in production from 2009 to 2010. The mandate was not sufficient to overcome the expiration of the tax credit, along with tight credit markets and an overall poor economy. The National Biodiesel Board maintains that the extension of the tax credit that was enacted in December 2010 will "help kick-start the domestic biofuel industry," and that the credits make biodiesel competitive with conventional diesel at the pump.[24]

While the tax incentives improve the competitiveness of BBD fuels, some stakeholders question whether they are warranted in the presence of a mandate to use the fuel. They argue that if the mandate is binding, the incentives will not promote the use of additional fuel beyond the mandate, and that overall fuel prices are lowered, creating an incentive to use more conventional fuel.[25] A counterargument is that this perverse incentive is only likely if the mandate is binding and credible. As seen in Figure 1, after the expiration of the tax credit, production declined from 2009 to 2010. If producers do not see the mandate as binding and certain, the presence of the tax credit may lower their perceived risk. Alternately, producers could be gambling on the likelihood that EPA will waive the BBD mandate in the face of limited production, or they could be hedging their bets into 2011 based on the deficit carryover contingency while they wait to see what the EPA will do with the BBD mandate. BBD producers have already seen the EPA waive the cellulosic RFS2 mandate in successive years due to lack of production. Meanwhile, BBD RIN prices have risen steadily to about $0.90 per gallon.

If BBD RIN prices remain high, producers may perceive less need for the tax credit, especially if RIN prices ultimately exceed the per-gallon value of the tax credit.

Implications for the Future

Going forward, the settlement of the 2009/2010 combined BBD mandate,[26] as well as the trajectory of domestic production/consumption and BBD RIN prices in 2011, could provide insights for the future of other biofuel/RIN markets. For the BBD RIN market to function effectively, EPA most likely will need to hold firm to its 2011 RFS2 BBD mandate. Because of shortfalls in cellulosic biofuel production capacity, EPA has waived the vast majority of the 2010 and 2011 cellulosic biofuel mandates.[27] EPA lowered the 2010 and 2011 mandates from 100 million gallons to 6.5 million gallons, and from 250 million gallons to 6.6 million gallons, respectively. In 2012, the scheduled mandate is 500 million gallons, and it remains unclear whether there will be sufficient production capacity to meet that level.

If cellulosic biofuel production capacity does eventually catch up with the RFS schedule, long-run cellulosic RIN prices may provide a significant financial incentive for cellulosic biofuel production, if the mandates are perceived as binding and credible. However, if EPA consistently provides waivers from the cellulosic biofuel mandate, and producers question the credibility of the mandate, cellulosic biofuel production—and cellulosic RIN prices—could stay low.

POTENTIAL POLICY OPTIONS

The markets for biomass-based diesel fuel and for RINs under the RFS raise many issues. Depending on one's perspective, there are a variety of policy options to address these issues. On the tax policy side, these options fall into three categories: (1) maintain the existing incentive structure and extend it into the future; (2) eliminate some or all of the incentives; (3) modify the incentives to more directly address certain policy goals.

In the first category—maintaining existing incentives—efforts to extend tax incentives have been proposed and enacted. Most recently, the "tax extenders" bill that was enacted in December 2010[28] extended through the end of 2011 various biofuels tax incentives, including the biodiesel and renewable diesel tax credits.[29] A key question for Congress is whether to extend these incentives past 2011 at their current levels, to modify them, or to let them expire.[30] Proponents of the tax incentives argue that they spur domestic production of non-fossil fuels, reducing U.S. reliance on imported petroleum, promoting rural development, and raising farm incomes.

As noted above, some critics argue that layering tax incentives on top of mandates is a redundancy, provides unintended incentives for conventional fuel supply, and may add little further incentive for production/consumption (beyond the mandate). Further, some stakeholders are concerned that as the RFS mandates grow, the liability for Treasury revenues will continue to increase if the tax incentives are maintained. Thus some stakeholders have argued for an elimination or phaseout of the tax incentives.

Other proposals have included modifying the existing tax incentives to base their value of performance measures, as opposed to simple per-gallon incentives. Possible performance measures could be based on various goals. For example, the value of the tax credit could be based on the expected amount of petroleum displaced by the fuel, or by its environmental performance (e.g., greenhouse gas emissions reduction). Further, one of the key rationales for the tax incentives is to make biofuels more competitive with petroleum. Thus, the tax incentives could be keyed to the price of oil: as oil prices rise above a certain level, the incentives could phase out.

One potential concern for BBD is that supply may not keep pace with mandates set in the RFS, regardless of the tax incentives. Therefore, some stakeholders may look for ways to make it easier to meet the BBD mandate, either through additional economic incentives or through a lowering of the mandate. The former would help spur production, while the latter would likely limit investment in new production.

End Notes

[1] For more information, see CRS Report R41282, *Agriculture-Based Biofuels: Overview and Emerging Issues*, by Randy Schnepf.

[2] For more information on biofuels incentives (both tax and non-tax), see CRS Report R40110, *Biofuels Incentives: A Summary of Federal Programs*, by Brent D. Yacobucci.

[3] For more information on the RFS, see CRS Report R40155, *Renewable Fuel Standard (RFS): Overview and Issues*, by Randy Schnepf and Brent D. Yacobucci.

[4] Advanced biofuels include ethanol from sugarcane, biomass-based diesel fuels, and ethanol and other fuels produced from cellulosic feedstocks.

[5] For more information on greenhouse gas reduction requirements under the RFS, see CRS Report R40460, *Calculation of Lifecycle Greenhouse Gas Emissions for the Renewable Fuel Standard (RFS)*, by Brent D. Yacobucci and Kelsi Bracmort.

[6] U.S. Department of Energy, Energy Information Administration, *Biodiesel Monthly Report*, Washington, DC, October 2010, Table 4, *http://www.eia.doe.gov/* cneaf/solar.renewables/ page/biodiesel/biodiesel.html.

[7] P.L. 108-357, §302.

[8] Biodiesel produced from virgin agricultural products (as opposed to recycled grease).

[9] For more information, see CRS Report R40110, *Biofuels Incentives: A Summary of Federal Programs*, by Brent D. Yacobucci.

[10] Energy is needed in the production of soybeans (and in the chemical inputs to soybean production), the transportation and crushing of soybeans into soybean oil, the conversion of soybean oil into biodiesel, and the transportation and distribution of the finished fuel. Various studies have calculated the fossil energy balance (the ratio of energy contained in the fuel to the sum of the fossil energy inputs), with a broad range of results which depend largely on assumptions about the energy value of various inputs and outputs to soy and biodiesel production. Recent analyses by the Department of Energy and EPA have found a large net energy balance, although some other studies have come to different conclusions. See Argonne National Laboratory, *Life-Cycle Assessment of Energy and Greenhouse Gas Effects of Soybean-Derived Biodiesel and Renewable Fuels*, March 12, 2008; U.S. Environmental Protection Agency, *Renewable Fuel Standard (RFS2) Regulatory Impact Analysis*, February 2010; and David Pimentel and Tad W. Patzek, "Ethanol Production Using Corn, Switchgrass, and Wood; Biodiesel Production Using Soybean and Sunflower," *Natural Resources Research*, vol. 14, no. 1, 2005, pp. 65-75.

[11] CRS analysis of data from Department of Energy, *Clean Cities Alternative Fuel Price Report*, Landover, MD, September 2005 to October 2010, http://www.afdc.energy.gov/afdc/ price_ report.html. USDA, Foreign Agricultural Service, *Oilseeds: World Markets and Trade* Monthly Circular, January 2011, http://www.fas.usda.gov/psdonline/ circulars/oilseeds.pdf.

[12] Within each batch, individual gallons of fuel are numbered sequentially. Thus, the last 16 digits of the 38-digit number indicate the total volume (and number of RINs) in that batch. For a more detailed discussion of RINs, see CRS Report R40155, *Renewable Fuel Standard (RFS): Overview and Issues*, by Randy Schnepf and Brent D. Yacobucci.

[13] The exception to this is that, because the 2009 program was established after the end of 2009, 2008 BBD RINs may be used to meet the 2009/2010 combined mandate.

[14] A cellulose-based BBD fuel could potentially be used to meet all four mandated categories.

[15] RINs are generally given an equivalency rating based on their energy content relative to ethanol. As the BBD mandate is for a given number of gallons of BBD specifically, the equivalency is one-to-one. But a biodiesel RIN is equivalent to 1.5 RINs of ethanol in meeting the advanced biofuel or overall RFS mandates.

[16] *Ethanol Monitor* and other publications currently track prices for conventional (corn-based) ethanol RINs and biodiesel RINs.

[17] EPA, "Regulation of Fuels and Fuel Additives: Changes to Renewable Fuel Standard Program," 75 *Federal Register* 14669-14904, March 26, 2010.

[18] EPA, *Questions and Answers on Changes to the Renewable Fuel Standard Program (RFS2)*, Washington, DC, October 13, 2010, *http://www.epa.gov/* otaq/fuels/renewablefuels/ compliancehelp/rfs2-aq.htm.

[19] EIA estimates that 176 million gallons of biodiesel was consumed from January through September 2010, leading to a monthly average of 19.6 million gallons. If that average continues through the fourth quarter of 2010, total consumption would be roughly 235 million gallons. EIA, *October 2010 Monthly Energy Review*, Washington, DC, October 28, 2010, Table 10.4, http://www.eia.doe.gov/mer/.

[20] This may be especially evident if fuel suppliers are expecting a shortfall in 2011 in the number of RINs available to meet the BBD mandate. As suppliers may not carry year-on-year deficits, suppliers may be less willing to register a deficit for the 2009/2010 mandate if they expect to need to carry a deficit from 2011 into 2012.

[21] EPA, "Regulation of Fuels and Fuel Additives: 2011 Renewable Fuel Standards," 75 *Federal Register* 76802, December 9, 2010.

[22] EIA, *Biodiesel Monthly Report*, Washington, DC, October 2010, Table 3, *http://www.eia.* doe.gov/ cneaf/ solar.renewables/page/biodiesel/biodiesel.html.

[23] According to the U.S. Department of Agriculture (USDA), U.S. production of soybean oil in 2009/2010 was 19.6 billion pounds. Mark Ash, *Oil Crops Outlook*, USDA, Economic Research Service, OCS-11a, Washington, DC, January 13, 2011, Table 3, *http://usda* .mannlib.cornell.edu/usda/current/OCS/OCS-01-13-2011.pdf.

[24] National Biodiesel Board, *Biodiesel Tax Incentive*, Jefferson City, MO, December 16, 2010, http://www.biodiesel.org/news/taxcredit/default.shtm.

[25] Harry de Gorter and David R. Just, *The Law of Unintended Consequences: How the U.S. Biofuel Tax Credit with a Mandate Subsidizes Oil Consumption and Has No Impact on Ethanol Consumption*, Cornell University, Charles H. Dyson School of Applied Economics and Management, Cornell University Working Paper No. 2007-20, Ithaca, NY, February 1, 2008, http://papers.ssrn.com/sol3/papers.cfm?abstract_id=1024525.

[26] February 28, 2011.

[27] See CRS Report R41106, *Meeting the Renewable Fuel Standard (RFS) Mandate for Cellulosic Biofuels: Questions and Answers*, by Kelsi Bracmort.

[28] The Tax Relief, Unemployment Insurance Reauthorization, and Job Creation Act of 2010 (P.L. 111-312)

[29] P.L. 111-312, §701. As the credit had expired at the end of 2009, the law made the extension retroactive to cover all 2010 production, as well.

[30] See CRS Report R40110, *Biofuels Incentives: A Summary of Federal Programs*, by Brent D. Yacobucci.

In: Biofuel Use in the U.S.
Editors: S. Alonso and M. R. Ortega
ISBN: 978-1-62100-441-7

Chapter 4

UNINTENDED CONSEQUENCES OF BIOFUELS PRODUCTION: THE EFFECTS OF LARGE-SCALE CROP CONVERSION ON WATER QUALITY AND QUANTITY

United States Geological Survey

In the search for renewable fuel alternatives, biofuels have gained strong political momentum. In the last decade, extensive mandates, policies, and subsidies have been adopted to foster the development of a biofuels industry in the United States. The Biofuels Initiative in the Mississippi Delta resulted in a 47-percent decrease in cotton acreage with a concurrent 288-percent increase in corn acreage in 2007. Because corn uses 80 percent more water for irrigation than cotton, and more nitrogen fertilizer is recommended for corn cultivation than for cotton, this widespread shift in crop type has implications for water quantity and water quality in the Delta. Increased water use for corn is accelerating water-level declines in the Mississippi River Valley alluvial aquifer at a time when conservation is being encouraged because of concerns about sustainability of the groundwater resource. Results from a mathematical model calibrated to existing conditions in the Delta indicate that increased fertilizer application on corn also likely will increase the extent of nitrate-nitrogen movement into the alluvial aquifer. Preliminary estimates based on surface-water modeling results indicate that higher application rates of nitrogen increase the nitrogen exported from the Yazoo River Basin to the

Mississippi River by about 7 percent. Thus, the shift from cotton to corn may further contribute to hypoxic (low dissolved oxygen) conditions in the Gulf of Mexico.

Top photograph: Furrow irrigation of corn acreage in the Mississippi Delta. Photograph courtesy of the Mississippi State University Extension Service.

Bottom photograph: Water-level monitoring of an alluvial aquifer well in Bolivar County, Mississippi. Photograph by Michael A. Manning, U.S. Geological Survey.

WHY HAS THE PRODUCTION OF BIOFUELS BECOME IMPORTANT?

Biofuels are fuels produced directly or indirectly from organic materials such as plants and animal waste. Corn-based ethanol is the most common type of biofuel produced in the United States—approximately 34 billion liters (L) were produced in 2008 (data accessed on June 7, 2010, at *http://www. afdc.energy*

Biofuels have received considerable support because they use renewable resources and have the potential to reduce greenhouse gas emissions. However, there are some environmental concerns attributed to the increase in corn acreage for biofuels production, such as the need for irrigation in some areas, the application of increased amounts of nitrogen fertilizers and water-soluble pesticides, and soil erosion from the tillage of crops (National Research Council of the National Academies, 2008).

The Biofuels Initiative (BFI), which was implemented by the U.S. Department of Energy (DOE) Biomass Program in late 2006, was developed by the DOE Office of Energy Efficiency and Renewable Energy to help meet the goals of the Energy Independence and Security Act (EISA; Public Law Number 110-140). The goal of the EISA is to increase the production of renewable and alternative fuels and reduce dependence on foreign oil within the United States. Two primary goals for the BFI and Biomass Program are to (1) produce 230 billion L of ethanol to replace 30 percent of current gasoline levels by 2030 and (2) reduce ethanol costs to prices that are competitive with gasoline by 2012.

THE MISSISSIPPI DELTA—CONSEQUENCES OF BIOFUELS PRODUCTION FROM A LOCAL PERSPECTIVE

The Yazoo River Basin is the largest river basin in Mississippi, with a drainage area of 34,590 square kilometers (km^2), and is divided equally between lowlands and highlands. An area referred to locally as the "Delta" covers about 18,130 km^2 of the lowlands part of the Yazoo River Basin in northwestern Mississippi. The U.S. Geological Survey's (USGS) National Water-Quality Assessment (NAWQA) Program has spent much of the last two decades studying the relation between agriculture and water quality and quantity in the Mississippi Delta (figure 1). Because of fertile soils and a long growing season, about 90 percent of the land is used for agriculture, including the cultivation of cotton, corn, rice, and soybeans. Although the climate is humid and subtropical, and the average rainfall ranges from 114 centimeters per year (cm/yr) in the north to 152 cm/yr in the south, about 28 percent of the rainfall occurs during the growing season (Snipes and others, 2005). Thus, irrigation is needed to maximize crop production.

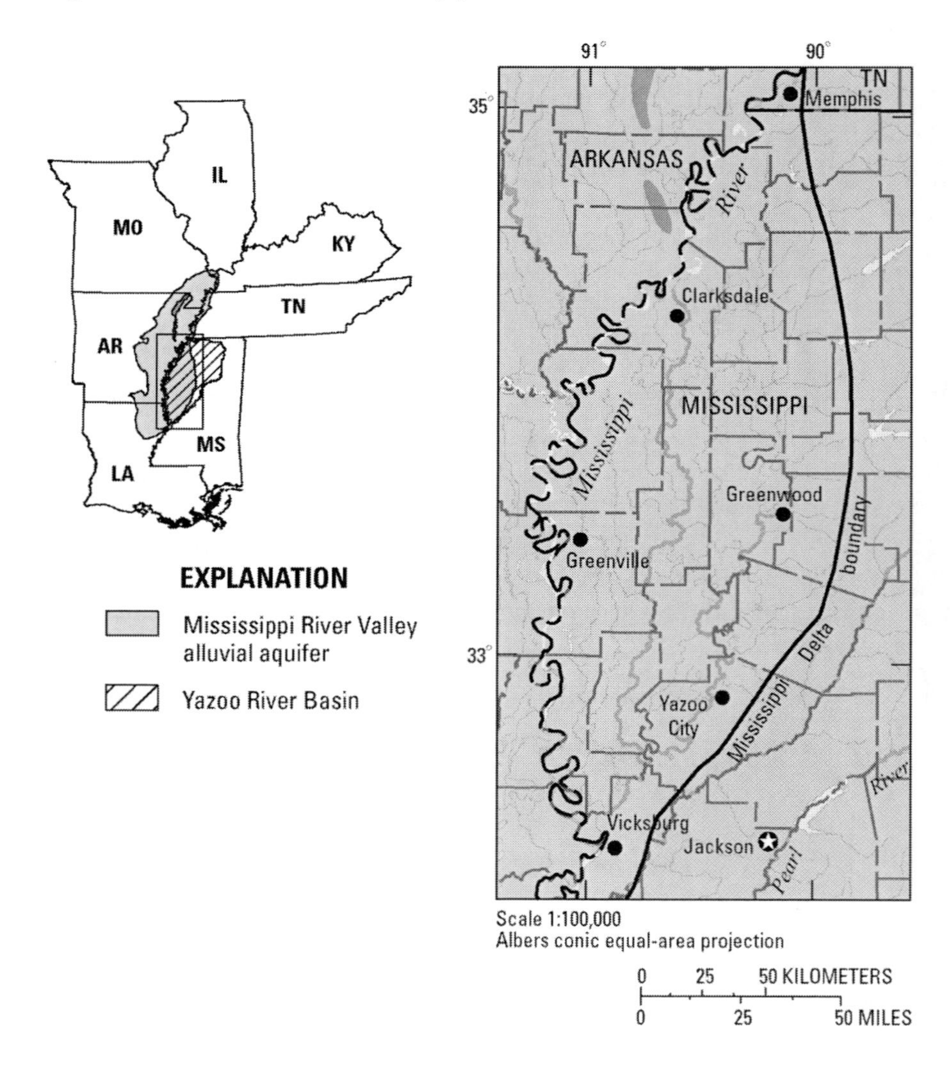

Figure 1. Areal extent of the Mississippi River Valley alluvial aquifer and the location of the Mississippi Delta in northwestern Mississippi.

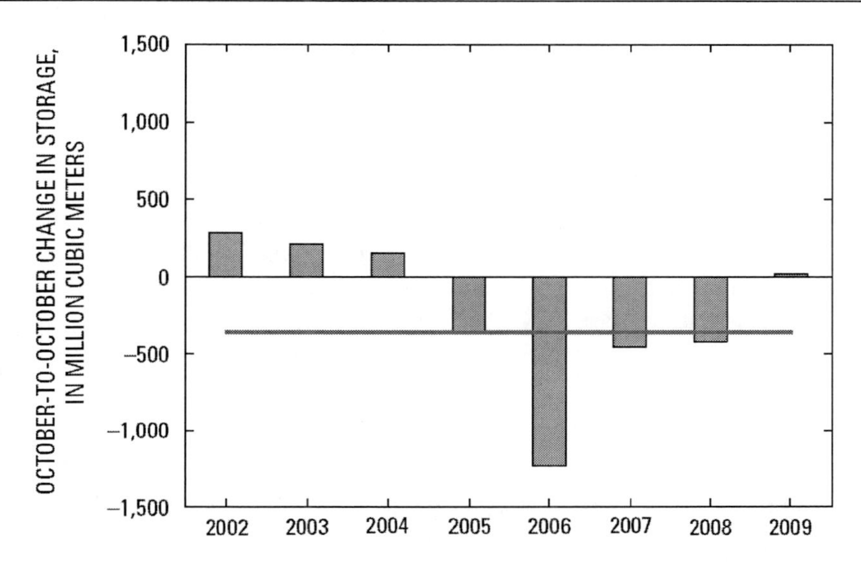

Figure 2. Annual (October-to-October) change in storage of the Mississippi River Valley alluvial aquifer in the Mississippi Delta, 2002–2009. Red line denotes the 22-year average change in storage for the aquifer (Mark Stiles, Yazoo Mississippi Delta Joint Water Management District, written commun., 2010).

The primary source of water for irrigation in the region is the Mississippi River Valley alluvial (MRVA) aquifer—a relatively thin hydrogeologic unit with an average thickness of 41 meters (m) and an areal extent of approximately 85,470 km^2. The aquifer is composed of highly permeable layers of sand, gravel, and silt (with minor amounts of clay) that underlie parts of Arkansas, Louisiana, Mississippi, Missouri, and to a lesser extent Illinois, Kentucky, and Tennessee (figure 1). In Mississippi, the aquifer is overlain by a relatively impervious clay layer, and therefore, only a small percentage of precipitation infiltrates the soils to directly recharge the aquifer. Estimates of recharge to the aquifer from precipitation range from 5.8 to 10.9 cm/yr (Arthur, 2001; Green and others, 2009; Welch and others, 2009). Because the Delta is a low, flat area with low permeability, the remaining precipitation is lost through evapotranspiration and to streamflow, which is supplemented by irrigation return flow during the summer months. During the winter, most of the runoff from rainfall drains into streams.

Approximately 15 million cubic meters per day (m^3/d) of water are withdrawn from the MRVA aquifer during the May to August growing season in Mississippi, which makes it the most heavily used aquifer in the State

(Maupin and Barber, 2005). Owing to the large amount of withdrawal, water levels in the aquifer have been declining since at least 1980. The average groundwater level has declined 0.1 meter per year (m/yr) or less in most areas, but has declined as much as 0.5 m/yr in some areas. Overall, the total volume of water stored in the aquifer has declined since 1980 (Arthur, 2001), and current withdrawals from the alluvial aquifer exceed recharge. From October 1987 to October 2009, there was an average annual loss in storage of approximately 355 million cubic meters (m^3) from the MRVA aquifer in the Mississippi Delta (Mark Stiles, Yazoo Mississippi Delta Joint Water Management District, written commun., 2010; figure 2).

Increased Withdrawals for Corn and Soybeans Are Accelerating Groundwater Declines in the MRVA Aquifer

Withdrawals from the MRVA aquifer have increased due to market-driven land conversion from cotton to corn and soybeans. The increase in corn prices driven by demand for ethanol-based biofuels resulted in a 47-percent reduction in cotton acreage concurrent with a 288-percent increase in corn acreage in 2007 relative to 2006 (figure 3). From 2007 to 2009, corn acreage decreased by 21 percent and soybean acreage increased by 46 percent in response to rising demand for soybeans to replace corn as feed for livestock (figure 3). Groundwater withdrawal rates (and application rates) for a particular crop vary from year to year, depending on weather conditions, irrigation methods, and other factors. For example, withdrawals for corn increased from 219 million m^3 in 2006 to 566 million m^3 in 2007, whereas withdrawals for cotton decreased from 855 to 282 million m^3 during this same period. Although corn acreage increased in 2007, groundwater application rates (withdrawal volume per unit land area in production) were about 30 percent lower in 2007 than in 2006 because of timely precipitation (figure 3; table 1). Because soybean production uses nearly as much water as corn production, the 2008 to 2009 shift in crop further increased the demand for groundwater from the MRVA aquifer.

Withdrawal rates for cotton production ranged from 0.09 million cubic meters per square kilometer per year ($m^3/km^2/yr$) in 2004 to 0.25 million $m^3/km^2/yr$ in 2006 (table 1). The average annual withdrawal rate from 2002 to 2009 was 0.15 million $m^3/km^2/yr$ for cotton, 0.22 million $m^3/km^2/yr$ for soybeans, and 0.27 million $m^3/km^2/yr$ for corn (table 1; Yazoo Mississippi Delta Joint Water Management District, 2009). Based on these averages, which normalize withdrawals for inter-annual differences in weather, converting 1 km^2 of cotton to 1 km^2 of corn results in a 0.12 million $m^3/km^2/yr$ increase in withdrawals from the MRVA aquifer. Corn production in the Delta increased from 598 km^2 in 2006 to 2,321 km^2 in 2007, while cotton production fell from 3,505 km^2 in 2006 to 1,853 km^2 during the same period (figure 3; http://www. nass.usda.gov/Statistics_by_State/Mississippi/index.asp; accessed April 10, 2010). Assuming that about 1,652 km^2 of cotton were converted to corn between 2006 and 2007, the additional increase in groundwater withdrawn, based on the measured 2007 application rates, was 165 million m^3. A similar analysis for the entire 2008 through 2009 period, including changes in soybean production, indicates that an additional 270 million m^3 of water were withdrawn from the MRVA aquifer from 2008 through 2009 as a result of crop conversions—a total of 435 million m^3 more groundwater withdrawn for the entire 2007 to 2009 period.

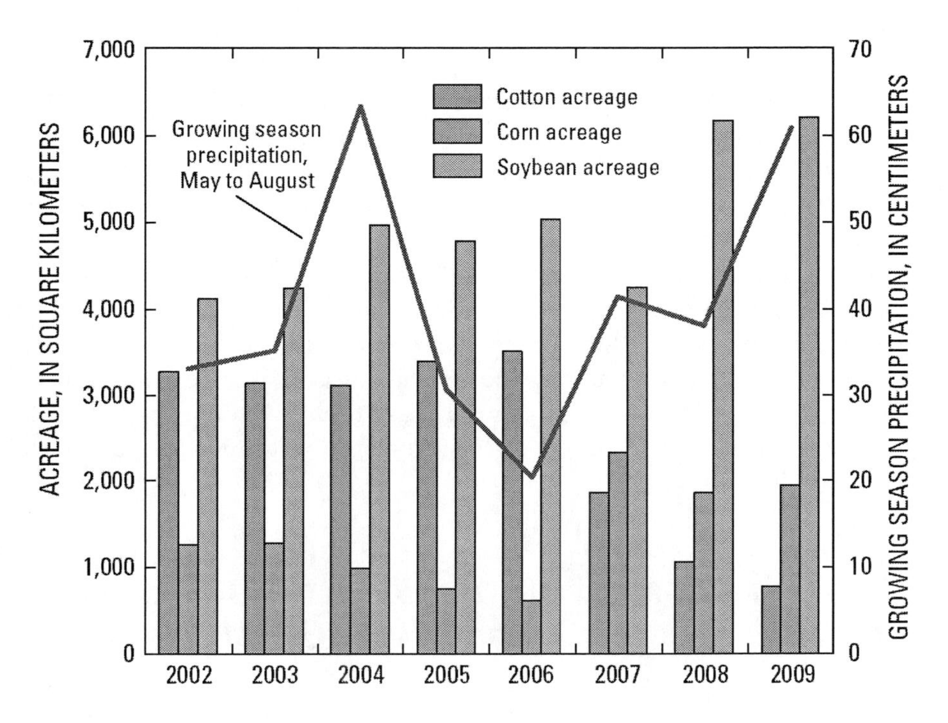

Figure 3. Crop acreage and total growing season precipitation in the Mississippi Delta, 2002–2009.

Table 1. Groundwater Withdrawal Rates for Cotton, Corn, and Soybean Acreage in the Mississippi Delta, 2002–2009

Year	Cotton	Corn	Soybeans
2002	0.15	0.28	0.22
2003	0.15	0.19	0.19
2004	0.09	0.12	0.12
2005	0.15	0.31	0.19
2006	0.25	0.37	0.31
2007	0.15	0.25	0.25
2008	0.19	0.37	0.31
2009	0.09	0.25	0.19
8-year average	0.15	0.27	0.22

Source: Yazoo Mississippi Delta Joint Water Management District (2009). All values are in million cubic meters per square kilometer per year.

Two scenarios for the volume of change in storage of the MRVA aquifer from 2007 to 2009 indicate that the conversion from cotton to corn and soybean acreage increased storage loss in the MRVA aquifer. A change in water withdrawals for cotton, corn, and soybeans was calculated for 2007 to 2009, using average withdrawal rates (table 1), in reference to 2006, the year prior to the large-scale crop conversion. In 2007, storage loss in the aquifer would have been approximately 250 million m^3 if the conversion from cotton to corn acreage had not occurred (figure 4). In 2008, storage loss in the aquifer would have been approximately 210 million m^3 if the conversion from cotton to corn and soybean acreage had not occurred (figure 4). A slight increase in storage of 18 million m^3 occurred in 2009 because of decreased groundwater withdrawals during the growing season as a result of timely rainfall. However, potential storage would have been 189 million m^3 if further increases in total acreage for soybean and corn, relative to 2006 levels, had not occurred (figure 4).

Increased Agrichemical Application Rates Have Affected Groundwater Quality

The USGS developed a mathematical advection-reaction water-quality model for a study site in the Delta to investigate changes in fertilizer application rates resulting from an increase in corn production and corresponding decrease in cotton production. The model was calibrated to existing conditions to assess fluxes of water and chemicals from agricultural fields to groundwater (Green and others, 2009; Welch and others, 2009). As with many other agricultural sites across the country, nitrate-nitrogen (N) contamination was found in shallow groundwater as a result of the leaching of chemical nitrogen (figure 5). At this site, nitrate-N was absent in samples at depths of 4.4 m below the water table and deeper due to a biodegradation reaction known as denitrification.

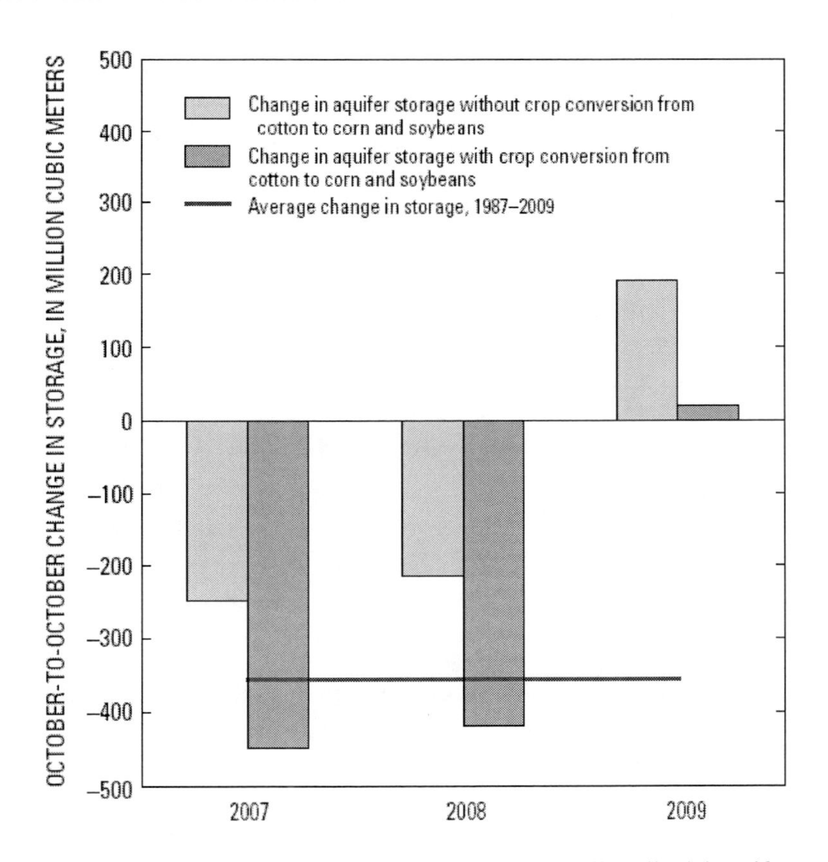

Figure 4. October-to-October change in Mississippi River Valley alluvial aquifer storage that can be attributed to the crop conversion from cotton to corn and soybeans, 2007–2009. For 2007, only the crop conversion from cotton to corn is considered.

The depth of leached nitrate-N was controlled largely by the slow downward movement of water, which is a function of soil properties, and the annual nitrogen fertilizer application rate to the overlying farm fields. Because of the link between fertilizer application rates and the degree of nitrogen contamination in ground-water, potential changes in nitrogen application rates in response to crop conversion for biofuels production have important implications for groundwater quality in the Mississippi Delta.

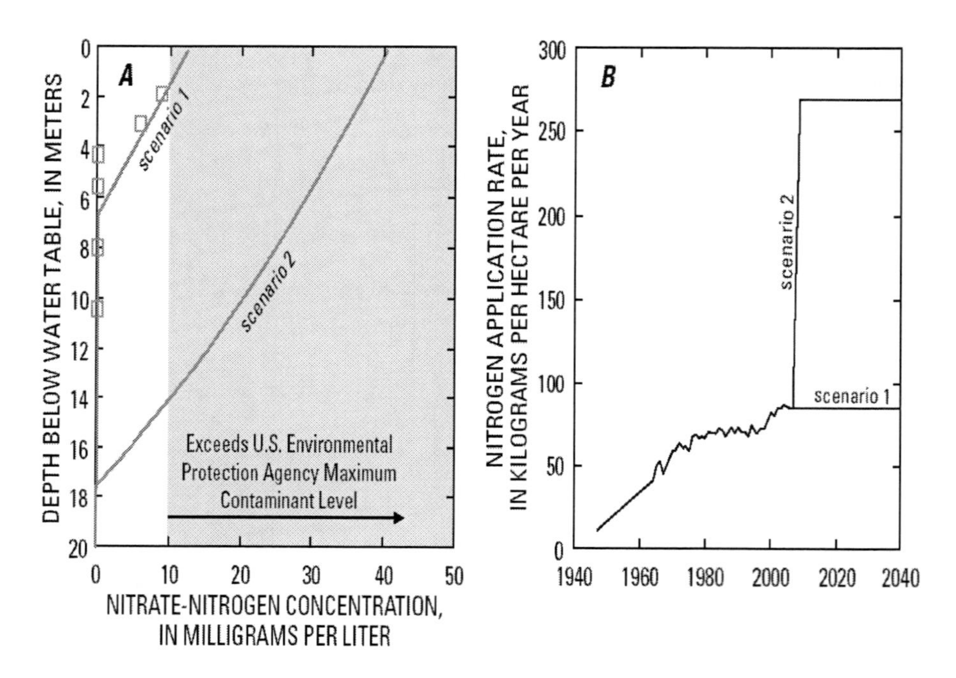

Figure 5. Two model scenarios for future agricultural practices and resulting water quality at the study site in the Mississippi Delta: (*A*) measured (indicated by green squares) and simulated nitrate-nitrogen concentrations in groundwater at various depths below the water table, and (*B*) fertilizer application for both scenarios over time. Scenario 1 assumes crop acreage and fertilizer application remain at 2006 levels, whereas scenario 2 is an immediate conversion to 100 percent of the recommended nitrogen application rate for corn.

The results of a mathematical model indicate that increased nitrogen fertilization due to additional corn production will expand the area of nitrate-N contamination of groundwater in Mississippi Delta groundwater. Figure 5 shows two hypothetical scenarios for future agriculture and water quality. In scenario 1, the crop types and fertilizer application in the Delta remain at 2006 levels (figure 5*B*). Nitrogen fertilizer application rates for this scenario are based on county agricultural chemical use from 1960 to 2008 (U.S.

Department of Agriculture, 2010). Under scenario 1, steady-state conditions are reached after 35 years, at which point predicted nitrate-N concentrations exceed the U.S. Environmental Protection Agency Maximum Contaminant Level (MCL) of 10 milligrams nitrate-N per liter (U.S. Environmental Protection Agency, 2006) in the upper 2 m of groundwater, and nitrate-N is transported to a maximum depth of 7 m below the water table (figure 5*A*). Once steady-state conditions are reached, the maximum depth of contamination remains at 7 m because the rate of downward transport reaches equilibrium with the rate of denitrification. In scenario 2, an increase in corn production results in a rapid increase in nitrogen application rates to 269 kilograms nitrate-N per hectare per year based on recommended fertilizer application rates from the Delta Planning Budget for corn acreage in Mississippi (figure 5*B*; Mississippi State University Department of Agricultural Economics, 2009). Under this scenario, steady-state conditions are reached after 95 years, at which point nitrate-N concentrations in groundwater are much higher than in scenario 1 and exceed the MCL up to 14 m below the water table. Nitrate-N migrates downward to a maximum depth of 18 m below the water table. Other hydrologic factors not considered in the model could result in an increase in the depth of migration of nitrate-N in the aquifer. For example, downward groundwater velocities could increase as a result of increased withdrawals.

NITROGEN EXPORT TO THE GULF OF MEXICO HAS INCREASED

Since at least 1980, a hypoxic zone has formed during the summer in the Gulf of Mexico off the coast of Louisiana. The zone has increased in size since measurements began in the early 1980s, and its extent is positively related to the annual amount of nitrogen entering the Gulf from the Mississippi River (Rabalais and others, 2002).

The USGS developed the *SPA*tially *R*eferenced *R*egressions on *W*atershed attributes (SPARROW) model to relate water-quality measurements to contaminant sources and environmental factors that affect both the rates of delivery of compounds to streams and their rates of in-stream processing. SPARROW has been used to identify the watersheds that are the largest contributors of nitrogen to the Gulf of Mexico, which includes the Yazoo River Basin (figure 6; Robertson and others, 2009).

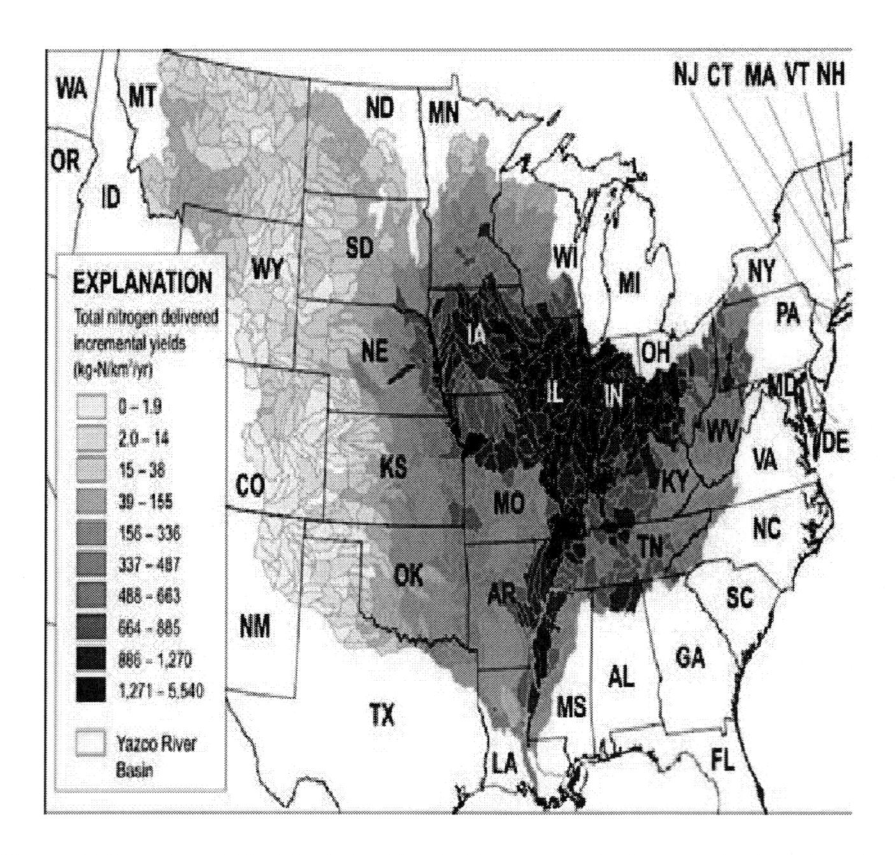

Figure 6. The extent of the Mississippi/Atchafalaya River Basin watersheds and total nitrogen delivered to the Gulf of Mexico from the basin. The Yazoo River Basin lies within the bold blue line (Robertson and others, 2009; kg-N/km2/yr, kilograms nitrate-nitrogen per square kilometer per year).

Using regional SPARROW model estimates of both the nitrogen application rates and the loss rates for nitrogen for the Yazoo River Basin, the conversion of cotton to corn acreage (comparing 2002 with 2007) is estimated to have caused a 7 percent increase in the nitrogen load for the Yazoo River (Richard A. Rebich, U.S. Geological Survey, written commun., 2010). Significant increases in nutrient inputs tied to changes in cropping patterns could mask any progress toward meeting the 45-percent nutrient reduction target by 2015 as specified by the Mississippi River/Gulf of Mexico Watershed Nutrient Task Force (2008).

STREAM ECOSYSTEM HEALTH IS AFFECTED BY DECLINING WATER QUALITY AND QUANTITY IN DELTA STREAMS

Groundwater input to a stream during periods of low rainfall is essential to stream ecosystem health because it maintains baseflow, regulates stream temperature, and generally is of higher quality than surface-water runoff. Prior to the rapid expansion of corn acreage in response to the BFI, baseflow in some Delta streams had been declining in recent years because of declining water levels in the MRVA aquifer (figure 7), and some stream reaches have remained dry (or nearly so) for months in the summer and fall during periods of low rainfall (figure 8).

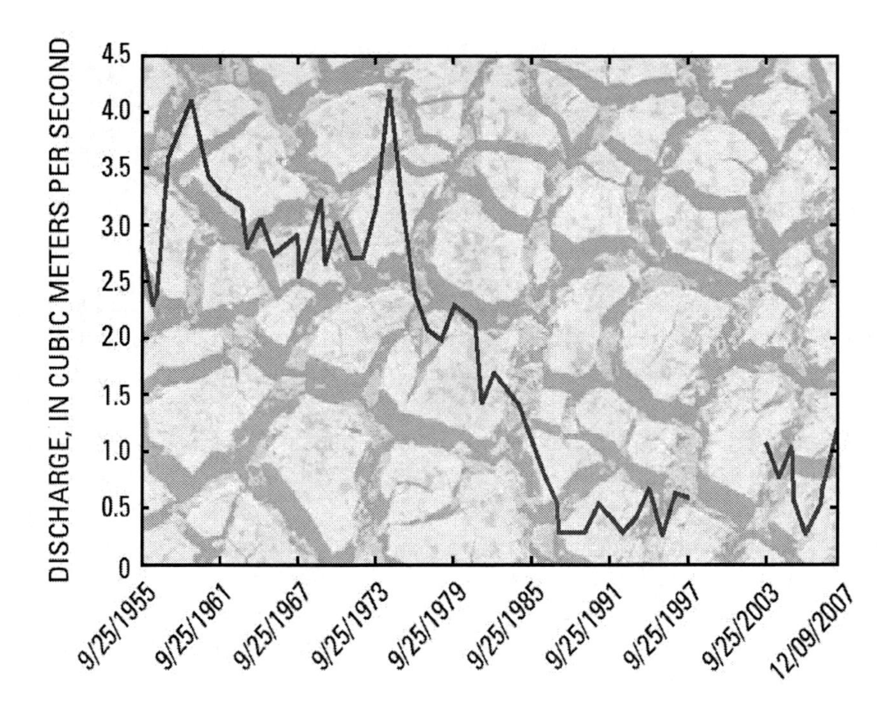

Figure 7. Annual minimum mean daily flow at the Big Sunflower River at Sunflower, Mississippi, showing decline in baseflow most evident since groundwater withdrawals from the Mississippi River Valley alluvial aquifer increased in the late 1980s. Data are missing for 1998 to 2002.

Figure 8. A Delta stream that is nearly dry during the summer because of loss of baseflow. Photograph by Matt Hicks, U.S. Geological Survey, September 25, 2007.

Most streams in the Delta have been hydrologically altered, and about 75 percent of the original forested wetlands have been cleared or drained, thus changing ecosystem structure and function (Creasman and others, 1992). A presumed decline in ecosystem health in Delta streams prior to 2007 is evidenced by a loss of habitat complexity and structure and a change in metabolic dynamics resulting from increased water temperatures and decreased dissolved oxygen levels (Kleiss and others, 2000). Continued declines in ecosystem health have led to further biodiversity losses through substantial changes in biological community assemblages and the increasing prevalence of more tolerant, but less desirable, species. The continued expansion of corn and soybean acreage in the Delta is expected to exacerbate the stresses on stream ecosystems primarily by (1) reducing groundwater inflow to streams—a change with particularly acute consequences during periods of low flow during the summer, and (2) increasing nutrient input to streams through increased fertilizer use.

SUMMARY: HOW HAS THE BIOFUELS INITIATIVE AFFECTED CONDITIONS IN THE MISSISSIPPI DELTA?

Higher groundwater withdrawals associated with the conversion of cotton to corn and soybean crops have exacerbated already declining water levels in the Mississippi River Valley alluvial (MRVA) aquifer leading to further loss of baseflow in most Delta streams.

Results from a mathematical water-quality model indicate that an increase of nitrogen fertilizer application to rates recommended for corn could increase nitrate-N contamination in much of the MRVA aquifer to levels that exceed the U.S. Environmental Protection Agency Maximum Contaminant Level. Maintaining fertilization rates at 2006 levels will result in only a minor increase in nitrate-N contamination.

Increased fertilizer application rates on corn have increased slightly (about 7 percent) the export of nitrogen from the Mississippi Delta to the Mississippi River and, ultimately, to the Gulf of Mexico where it may further exacerbate the hypoxic zone.

REFERENCES

[1] Arthur, J. K. (2001). *Hydrogeology, model description, and flow analysis of the Mississippi River alluvial aquifer in northwestern Mississippi*: U.S. Geological Survey Water-Resources Investigations Report 01–4035, 47 p.

[2] Creasman, L., Craig, N. J. & Swan, M. (1992). *The forested wetlands of the Mississippi River—An ecosystem in crisis*: Baton Rouge, Louisiana, The Nature Conservancy, 23 p.

[3] Green, C. T., Welch, Heather & Coupe, Richard. (2009). Multi-tracer analysis of vertical nitrate fluxes in the Mississippi River Valley alluvial aquifer [abs.], *in Eos Transactions of the American Geophysical Union,* v. *90*, no. 52, H31C-0799.

[4] Kleiss, B. A., Coupe, R. H., Gonthier, G. J. & Justus, B. G. (2000). *Water quality in the Mississippi Embayment, Mississippi, Louisiana, Arkansas, Missouri, Tennessee, and Kentucky*, 1995–1998: U.S. Geological Survey Circular 1208, 36 p.

[5] Maupin, M. A. & Barber, N. L. (2005). *Estimated withdrawals from the principal aquifers in the United States, 2000*: U.S. Geological Survey Circular 1279, 46 p.

[6] Mississippi River/Gulf of Mexico Watershed Nutrient Task Force. (2008). Gulf hypoxia action plan 2008 for reducing, mitigating, and controlling hypoxia in the northern Gulf of Mexico and improving water quality in the Mississipi River Basin, accessed May 5, 2010, at http://www.epa.gov/owow_keep/msbasin/pdf/ghap2008_ update082608.pdf.

[7] Mississippi State University Department of Agricultural Economics. (2009). *Delta 2010 planning budgets*: Budget Report 2009–06, 185 p.

[8] National Research Council of the National Academies. (2008). *Water implications of biofuels production in the United States:* Washington, DC, The National Academies Press, 61 p.

[9] Rabalais, N. N., Turner, R. E. & Wiseman, W. J., Jr. (2002). Gulf of Mexico hypoxia, a.k.a. "The Dead Zone": *Annual Review of Ecological Systems*, v. *33*, p. 235–263.

[10] Robertson, D. M., Schwarz, G. E., Saad, D. A. & Alexander, R. B. (2009). Incorporating uncertainty into the ranking of SPARROW model nutrient yields from Mississippi/Atchafalaya River Basin watersheds: *Journal of the American Water Resources Association*, v. *45*, no. 2, p. 534–549.

[11] Snipes, C. E., Nichols, S. P., Poston, D. H., Walker, T. W., Evans, L. P. & Robinson, H. R. (2005). *Current agricultural practices of the Mississippi Delta: Mississippi Agricultural and Forestry Experiment* Station Bulletin 1143, 18 p.

[12] U.S. Department of Agriculture. (2010). Economic Research Service Fertilizer Consumption and Use, accessed January 29, 2010, at http://www.ers.usda.gov/Data/FertilizerUse/.

[13] U.S. Environmental Protection Agency. (2006). 2006 edition of the drinking water standards and health advisories: U.S. Environmental Protection Agency, EPA-822-R-06-013.

[14] Welch, H. L., Green, C. T. & Coupe, R. H. (2009). The fate and transport of nitrate through the unsaturated zone at a site in northwestern Mississippi [abs.], *in* Geological Society of America 2009 Annual Meeting: Geological Society of America Abstracts with Programs, v. 41, no. 7, p. 29.

[15] Yazoo Mississippi Delta Joint Water Management District. (2009). Irrigation water use in the Mississippi Delta 2009 report, accessed April 10, 2010, atwateruse/2009%20Mississippi%20Delta%20Irrigation%20 Water%20Use%20Report.pdf.

ADDITIONAL INFORMATION

For additional information about the Biofuels Initiative and (or) the Mississippi Delta contact:

U.S. Geological Survey
Mississippi Water Science Center
308 South Airport Road
Jackson, MS 39208-6649 http://ms.water

CHAPTER SOURCES

Chapter 1 -This is an edited, reformatted and augmented version of the United States Department of Agriculture, Economic Research Service, *Effects of Increased Biofuels on the U.S. Economy in 2022,* Report Number 102, dated October 2010.

Chapter 2 -This is an edited, reformatted and augmented version of United States Government Accountability Office, Report to Congressional Requesters, GAO-11-513, *Challenges to the Transportation, Sale, and Use of Intermediate Ethanol Blends,* dated June 2011.

Chapter 3 -This is an edited, reformatted and augmented version of Congressional Research Service, *The Market for Biomass-Based Diesel Fuel in the Renewable Fuel Standards (RFS),* Report R41631, dated February 11, 2011.

Chapter 4 -This is an edited, reformatted and augmented version of United States Geological Survey, *Unintended Consequences of Biofuels Production,* Open-File Report dated 2010-1229.

Index

D

E

M

N

O

P

Q

R